Dr Rhodri Evans studied physics at Imperial College London, graduating with first-class honours, before gaining his PhD in astrophysics from Cardiff University. He has taught at the University of Toledo in the USA, at Swarthmore College and done post-doctoral research at the University of Chicago's Yerkes Observatory, the birthplace of modern astrophysics. He is currently a research fellow at Cardiff University with a particular interest in infrared astronomy. Rhodri is the author of numerous academic papers as well as articles on popular science, he speaks regularly at conferences and is a frequent contributor to the BBC on physics and astronomy. His popular blog can be found at thecuriousastronomer.wordpress.com.

Brian Clegg is one of the UK's leading science writers. His books have been selected as *The Times*'s book of the week, won the IVCA Clarion award and been translated into many languages. He read natural sciences, specialising in experimental physics, at Cambridge University. At British Airways he formed a new department tasked with developing high-tech solutions for the airline before setting up his own creativity consultancy, running courses on the development of ideas and the solution of business problems for many high-profile clients. Brian has written for numerous magazines and newspapers, including the *Observer*, the *Wall Street Journal*, *Nature*, *BBC Focus*, *Physics World* and *The Times*. He is a regularly contributor on radio and TV and a popular speaker in schools, universities and at the Cheltenham Festival of Science. Brian is the editor of the successful www.popularscience.co.uk book review site and blogs at www.brian-clegg.blogspot.com.

T018699ᄅ

Ten Physicists

WHO TRANSFORMED OUR
UNDERSTANDING OF REALITY

....

RHODRI EVANS

AND

BRIAN CLEGG

RUNNING PRESS
PHILADELPHIA · LONDON

ROBINSON

First published in Great Britain
in 2015 by Robinson

Copyright © Rhodri Evans
and Brian Clegg 2015

3 5 7 9 8 6 4

The moral right of the authors
has been asserted.

All rights reserved.
No part of this publication may be
reproduced, stored in a retrieval system,
or transmitted, in any form, or by any
means, without the prior permission
in writing of the publisher, nor be
otherwise circulated in any form of
binding or cover other than that in which
it is published and without a similar
condition including this condition being
imposed on the subsequent purchaser.

A CIP catalogue record for this book
is available from the British Library.

ISBN 978-1-47212-037-3 (paperback)
ISBN 978-1-47212-038-0 (ebook)

Typeset by Hewer Text UK Ltd, Edinburgh
Printed and bound in Great
Britain Clays Ltd, St Ives plc

Papers used by Robinson are
from well-managed forests and
other responsible sources

MIX
Paper from
responsible sources
FSC FSC® C104740
www.fsc.org

Robinson
An imprint of
Little, Brown Book Group
Carmelite House
50 Victoria Embankment
London EC4Y 0DZ

An Hachette UK Company
www.hachette.co.uk

www.littlebrown.co.uk

First published in the United States in
2015 by Running Press Book Publishers,
A Member of the Perseus Books Group

All rights reserved under the Pan-American
and International Copyright Conventions

This book may not be reproduced in whole or in
part, in any form or by any means, electronic
or mechanical, including photocopy, recording,
or by any information storage and retrieval
system now known or hereafter invented,
without permission from the publishers.

Books published by Running Press are
available at special discounts for bulk
purchases in the United States
by corporations, institutions
and other organizations.

For more information, please contact
the Special Markets Department at the
Perseus Books Group, 2300 Chestnut
Street, Suite 200, Philadelphia, PA 19103,
or call (800) 810-4145, ext. 5000, or email
special.markets@perseusbooks.com.

US ISBN: 978-0-7624-5812-7
US Library of Congress Control
Number: 2015945702

9 8 7 6 5 4 3 2 1
Digit on the right indicates the
number of this printing

Running Press Book Publishers
2300 Chestnut Street
Philadelphia, PA 19103-4371

Visit us on the web!
www.runningpress.com

Picture credits: p. 6, Gallileo: Science Photo
Library; p. 31, Isaac Newton: © National
Portrait Gallery, London; p. 49, Newton's
Principia Mathematica: Science Museum/
Science & Society Picture Library; p. 58,
Michael Faraday: © National Portrait
Gallery, London; p. 73 Faraday's Disc:
Library of Congress/Science Photo Library;
p. 80, James Clerk Maxwell: Emilio Segre
Visual Archives/American Institute of
Physics/Science Photo Library; p. 110,
Marie Curie: W.F. Meggers Collection/
American Institute of Physics/Science Photo
Library; p. 137, Radium Institute, Paris:
Universal Images Group/Getty images;
p. 140, Ernest Rutherford: © Peter Lofts
Photography/National Portrait Gallery,
London; p. 152, Albert Einstein: Associated
Press/Science & Society Picture Library;
p. 176, Niels Bohr: Science Museum/
Science & Society Picture Library; p. 198.
Paul Dirac: © Peter Lofts Photography/
National Portrait Gallery, London; p. 213:
positron: Carl Anderson/Science Photo
Library; p. 222, Richard Feynman: Science
Museum/Science & Society Picture Library

For our respective families
Maggie, Meirin, Siân-Azilis and Esyllt
Gillian, Chelsea and Rebecca

ACKNOWLEDGEMENTS

We would like to thank Duncan Proudfoot and all at Robinson for their support and Steven Weinberg for taking the time to help us.

Contents

The Ten

Foreword

....

STEVEN WEINBERG

It's a pretty good list. Perhaps it's a little Anglocentric. No one would dream of dropping Newton or Maxwell from a list of the ten greatest physicists and I would want to keep Rutherford and Faraday, but why Dirac and not Werner Heisenberg or Erwin Schrödinger? I would drop two names from the current list (no, I won't say which), keep Dirac, and add Heisenberg and Schrödinger. I would also add Christiaan Huygens and Ludwig Boltzmann. Yes, I know that makes twelve, but isn't physics important enough to warrant a dozen names in a list of the top ten? My wife and I occasionally play the game of listing the top ten movies of all time and we often find we include about a hundred.

Any list of this kind prompts a reflection on the difference between science and the arts. Science is a cumulative enterprise that leaves no space for the pioneers. We judge past scientists according to their contribution to our present understanding. For over a millennium, when natural philosophers said 'physics' they meant the physics of Aristotle. But our present physics owes nothing to Aristotle (rather the reverse) and it would be crazy to include him in a list of the top ten. In contrast, we admire J. M. W. Turner not because he foreshadowed impressionism but because he painted beautiful pictures and it would not be at all strange to include Homer or Sappho in a list of the ten greatest poets. In a list of the ten (or twelve) greatest physicists, we can trace the history of our progress in explaining the world.

Steven Weinberg is an American theoretical physicist who, with Sheldon Glashow and Abdus Salam, was awarded the 1979 Nobel Prize in physics for his contribution to the theory of the unified weak and electromagnetic interaction between elementary particles. Weinberg received a degree from Cornell and a PhD at Princeton. He has since worked at Columbia, Berkeley, MIT, Harvard and most recently the University of Texas at Austin.

Introduction

There is nothing we like better than a list. Newspapers and TV channels find they make great, cheap copy that is always bound to generate interest. The media bombard us with the hundred best pieces of classical music, the twenty books to read before you die or the top ten restaurants. Readers and viewers love them too. The appeal of a heady mix of recommendation and potential for dispute is hard to resist. We've all thought, Why did they select that one? Why did they leave out that one?

For lists about a commodity it's easy enough to try the different options and see if you agree, but there can be more subtlety when it comes to people. There is no problem putting together a rich list, but it is very different when considering intellectual achievement. How can you possibly pick a definitive top ten from the world's physicists throughout history, as was done in 2013 for the *Observer* newspaper? Certain figures would be hard to ignore – Newton and Einstein spring to mind – but there are plenty at the slightly-less-famous level who vie for the remaining places.

To make matters worse, the greatest physicists are not always the most famous. Some might be surprised, for instance, to find Stephen Hawking missing. He is, without doubt, the best-recognised living physicist. The chances are that a public vote would put him high in the top ten – and yet he doesn't appear in our list. This isn't because his work is considered unimportant, but there are a whole host of other physicists who didn't make the cut who would be placed above Hawking by anyone who knows the field.

A useful illustration of the tensions involved in putting such a list together was in the question we heard as we developed this book: 'Are you going to include Tesla?' The answer was easy: no. Nikola Tesla was not on the original list and it is hard to see how he could have been because he wasn't a physicist. Tesla began as an outstanding electrical engineer who made huge steps forward in the development of AC current and invented excellent AC motors, high-voltage generators, radio-controlled devices and more. But he had little grasp of twentieth-century physics. The point is that while some individuals have a place in the popular imagination, it doesn't follow that they are worthy of joining such a prestigious list.

Who, then, makes the grade? Let's take another look at the full list, published in the *Observer* on 12 May 2013 in an article by Robin McKie, science and technology editor for the *Observer*:

1   Isaac Newton (1643–1727)
2   Niels Bohr (1885–1962)
3   Galileo Galilei (1564–1642)
4   Albert Einstein (1879–1955)
5   James Clerk Maxwell (1831–1879)
6   Michael Faraday (1791–1867)
7   Marie Curie (1867–1934)
8   Richard Feynman (1918–1988)
9   Ernest Rutherford (1871–1937)
10 Paul Dirac (1902–1984)

To better understand why these individuals are here we need to have a feel for what physics is and how it fits into the wider world of science. It's possible to think of science as a pyramid. Physics (hand-in-hand with mathematics) forms the base – the fundamentals, the building blocks on which everything else is constructed. Chemistry takes the physics of atoms and molecules and studies their behaviour on a larger scale. And biology combines the working levels of both chemistry and physics in the study of the uniquely complex phenomenon that is life.

The men and woman on this list gave us real advances in our understanding of those fundamentals. Without physics – without the work of these physicists (and many others) – science as we know it really couldn't exist. Nor could the technology that is essential to the modern world. Science had only a small part to play in industry until the nineteenth century but, with the introduction of mechanisation, physics came to the fore – and it is still at the heart of everything from the sophisticated electronics of a smartphone to the simple workings of a fridge.

As Steven Weinberg makes clear in his foreword, this list is not the only possible selection. McKie made some interesting decisions. The most controversial aspect is placing Niels Bohr in the second spot. Few would argue about the vast contributions of Galileo, Newton and Einstein. But Bohr's work was more subtle. He gave us the first workable model of the atom and was the lead architect (if not the biggest contributor) in the development of quantum theory. But second place in the whole of history? Really?

It is interesting to look at the comments that accompany the original article. Leaving aside the surprised reactions to the omission of Tesla, many point out the lack of other big names from the twentieth century, particularly the founders of quantum physics. And there's an impassioned plea – not without reason – for Archimedes to be included. We will come back to the order of the list and who should or shouldn't be on it in the concluding chapter.

One guide to potential candidates since the early twentieth century tends to be the Nobel Prize. When the Swedish-born inventor and explosives tycoon Alfred Nobel left the majority of his estate to fund prizes for those who 'during the previous year, shall have conferred the greatest benefit on mankind', he not only shocked and hurt his family, but started a new mechanism for flagging up important developments in a number of fields, notably – for us – in physics.

While the earlier entries in our list have to be made without consideration of the Nobels, the physics prize was first awarded in 1901 (to Wilhelm Röntgen for the discovery of X-rays) and since then it has provided a marker of excellence. There are, of course, issues. The prize is limited to a maximum of three recipients who must be alive when the prize is awarded. This causes problems for the increasingly large teams involved in scientific research – and the often significant lag between work and prize also means that some potential recipients are no longer alive by the time the ceremony takes place.

In addition, while many famous characters in physics have taken their place among Nobel laureates, glance through the list of winners and you will see names that few would recognise. Ask anyone, physicists included, for the significance of the work of Nils Gustaf Dalén, and you will get a blank look. This is not entirely surprising as he won the prize in 1912 for his invention of 'automatic regulators for use in conjunction with gas accumulators for illuminating lighthouses and buoys'. There is no doubt that Tesla deserved a Nobel more than Dalén, but the ventures of the physics committee into technology are always fraught. (In lasers, for instance, the three winners do not include the patent holder or the man who constructed the first working laser.) Even in pure physics there can be controversies, as when the discoverer of the pulsar, Jocelyn Bell, was omitted while her boss got the prize. But most would agree that the Nobels provide a good starting point for finding modern members of a top ten list.

Our initial inclination was to explore the list from the *Observer* in reverse order, working up to presenting the gold to Isaac Newton. But that presents a problem. It would have meant starting off with Paul Dirac. Yet Dirac's work built on everything that came before him. So if we are sensibly to explore why these individuals achieved this accolade, it makes more sense to look at the ten physicists in chronological order.

This means that we begin our exploration of the list with a name that is as familiar as that of Newton. A name that has strong associations in the imagination, whether it is for dropping balls off the tower of Pisa or rebelling against the Church's insistence that the Earth was at the centre of the universe. That name is, of course, Galileo Galilei.

....

CHAPTER ONE

Galileo Galilei

....

It is no exaggeration to call Galileo the founder of modern physics, or even modern science. When Galileo was born, it was customary for 'natural philosophers' (as scientists were called) to follow the teachings of the ancient Greek philosopher Aristotle. The Italian poet Dante referred to Aristotle as 'the Master of those who know'; he was held in such high regard that virtually no one questioned his ideas in detail. Until Galileo came along.

Central to Aristotelian philosophy was the aim of understanding *why* things are the way we find them. Aristotle argued that it was necessary to grasp the ultimate purpose behind events in nature. Following his predecessor Empedocles, he believed that the world was composed of four elements – earth, air, water and fire. Beyond our world lay the celestial sphere, comprising the Moon, Sun, planets and stars. This was the realm of a fifth element, 'quintessence', and Aristotle taught that the heavens were unchanging and perfect.

Associated with the terrestrial elements were four 'qualities' in paired opposites – heat and cold and moisture and dryness. The four elements belonged in natural places and, through either gravity or levity (gravity's supposed natural opposition, a tendency to float upwards), the elements would try to return to their natural place. Aristotle laid out logical rules through which causes could be determined for natural events, causes that were arrived at by reason rather than through experimentation. Experiments were dismissed because they relied on fallible senses; instead, the laws governing nature were determined by thought alone. Galileo would challenge this accepted world order when evidence countered accepted wisdom.

Galileo Galilei was born in Pisa on 15 February 1564. His father Vincenzo was a professional lute player and musical theorist, who married Galileo's mother Giulia Ammannati in 1562. Galileo, named after a distant relative, the doctor Galileo Bonaiuti, was the eldest of seven (or possibly eight) children, and as befitted a child in a middle-class family, he was educated privately. When he was about ten, his family moved from Pisa to Florence, where Galileo continued his schooling before being sent to the Camaldolese monastery at Vallambroso. Galileo told his family that he wished to train to be a priest, but his father wanted Galileo to follow in the footsteps of his namesake and become a doctor. Vincenzo returned Galileo to Florence where he continued his religious studies through correspondence.

In 1581, at the age of seventeen, Galileo went to university in Pisa. This was a relatively late age – students typically started at thirteen or fourteen. It didn't take long for Galileo to realise that he was more interested in science than the priesthood, in part thanks to the instruction of one of Italy's most renowned natural philosophers, Andrea Cesalpino. Hearing his lectures may have been instrumental in Galileo switching allegiance from medicine to mathematics.

Galileo had a keen eye, making him naturally curious about the phenomena he saw around him. There are many stories surrounding Galileo's life and work and it is difficult to know how many actually happened; by the end of his life he had become such a great figure that stories were invented to perpetuate and increase his legend. For example, the account of his dropping balls of different weights from the leaning tower of Pisa to see whether they fell at the same rate is unlikely to be true. Another such story with uncertain origins concerned his observations of chandeliers in the cathedral in Pisa. Galileo was supposed to have been sitting through a particularly boring sermon when he noticed that the time it took for the chandeliers in the cathedral to swing back and forth seemed to depend on the length of their chains. Those on longer chains took more time than the chandeliers with shorter chains.

What is undisputed is that, in the summer after his first year at university, Galileo immersed himself in experiments to investigate the properties of pendulums. He had seen how his father experimented with musical instruments, carefully taking notes, and only altering one aspect of the experiment at a time. Galileo constructed pendulums with bobs of different weights and lengths of strings, setting them in motion with swings of different sizes and timing their motion with his pulse. He concluded that the period of a pendulum depends only on the length of the string and not on the size of the swing or the weight of the bob. (In fact, Galileo was wrong about the period's independence from the size of the swing – his observation only holds for small displacements.)

Galileo never made direct use of this discovery, though he would use the observations to help show that Aristotle was wrong in saying weights of different size fell at different speeds. However, Galileo's pendulum work later formed the basis of the development of the pendulum clock by Dutch scientist Christiaan Huygens in the seventeenth century.

Although Galileo was clearly bright as a student, he was also argumentative and was failing most of his courses, except mathematics. Still, he plodded on with his medical studies until his teachers at Pisa recommended that he switch to studying mathematics full time and Vincenzo reluctantly agreed. He knew mathematicians were no better paid that lute players but realised that his son's true talents were not in medicine.

Galileo continued for some time at Pisa, but left in 1585 without completing his degree. This was mainly due to Vincenzo suffering a reversal of fortunes and needing Galileo to come home to help support the family. Not receiving a degree was not uncommon for gentlemen of Galileo's background: university was a kind of finishing school for middle-class men and graduating was not deemed particularly important. It was the experience of attending university and the connections made there that mattered.

Galileo set about preparing himself for the role of mathematics professor. He took on tutoring jobs in Florence and Siena and towards the end of 1587 discovered an ingenious method for determining the centres of gravity of some solids. This earned him his first recognition beyond Italy and on the strength of this work Galileo applied unsuccessfully for a vacant professorship at the prestigious University of Bologna in 1588.

Although he had failed to secure the position, Galileo's work aroused the interest of the Marquis Guidobaldo del Monte, a powerful man, who remained one of Galileo's patrons until his death in 1607. The discovery also led to Galileo coming to the attention of Christopher Clavius, a mathematician and astronomer at the Jesuit college in Rome. With the help of del Monte and Clavius, Galileo was given a lecturing position at the University of Pisa in 1589, four years after leaving as a student. It was poorly paid but having an academic position made it possible for Galileo's patrons to work on securing him a better situation at Padua, where mathematics was highly regarded.

In 1590, Galileo wrote his manuscript *De Motu* (on motion). A discussion of Aristotelian ideas about the motion of bodies, it mixes his philosophy with mathematical ideas from Archimedes, a hero of Galileo's. *De Motu* does not contain much original work and it is clear from his discussion of Greco-Egyptian astronomer Claudius Ptolemy's *Syntaxis Mathematica* (generally known by its Arabic name, the *Almagest*) that at this time Galileo accepted that the Earth was at the centre of the universe.

It is in *De Motu* that Galileo first outlines his arguments that objects with different weights fall at the same rate. Aristotelian philosophy argued that objects fall if they have a preponderance of the elements water and earth, both of which want to find their natural place at the centre of the universe. Heavier objects contained a greater amount of such elements and so fell faster than lighter ones. Amazingly, this argument had never actually been verified but merely passed on as a given. Despite *De Motu* being more complete than any other work on motion at the time, Galileo never published it. This is probably because he was not happy with ideas he had on the motion of objects on inclined planes, that had yet to be experimentally verified.

Galileo's three-year lecturing position at Pisa was drawing to an end and, although he had become close friends with a few of his colleagues, he had succeeded in antagonising many of the professors. He had every reason to think his employment would not be renewed. To make matters worse, Vincenzo died in 1591 and Galileo, as the eldest son, became responsible for paying his eldest sister Virginia's dowry. Luckily, a better-paid position became available at Padua, which Galileo secured with the help of del Monte and Clavius. He took the post in 1592, receiving three times his previous salary.

In addition to the academic life at the University, Padua was a city with a thriving intellectual community. Much of this was centred on the home of G. V. Pinelli, who regularly invited people for discussions. When Galileo first arrived in Padua he briefly stayed with Pinelli and it was probably there that he met friar Paolo Sarpi and Cardinal Robert Bellarmine, each of whom would play an important part in Galileo's life.

There is no evidence of Galileo showing any interest in astronomy until 1595, the year he started thinking about an explanation for the Earth's tides. For many centuries, scientists had struggled to explain the two high tides and two low tides a day and the way that the timing of those tides changes. The (incorrect) explanation that Galileo developed required that the Earth both rotate on its axis and orbit the Sun. This was the first indication of Galileo toying with the heliocentric model that Copernicus proposed in 1543. This idea of the universe was opposed both by the church and natural philosophers, whose Aristotelian dogma required the Earth to be the centre of everything so that heavy objects would be drawn to it. In addition, the idea that the Earth was moving rapidly seemed preposterous. How could we not feel such motion? As things stood at the end of the sixteenth century, few accepted the Copernican model. The best data available in the late 1500s had come from Danish astronomer Tycho Brahe, but his work actually agreed more with a geocentric model rather than the heliocentric one. Brahe had proposed a hybrid model with the Sun and Moon orbiting a fixed Earth while the planets moved around the Sun.

In 1597, a visitor gave Galileo a copy of Johannes Kepler's book *Mysterium Cosmographicum* (the cosmographic mystery), published in 1596. This book was strongly pro-Copernican and must have influenced Galileo's thinking. Galileo wrote to Kepler stating that he had long been a supporter of the 'new astronomy' and that by using it he was able to explain some things that could not otherwise be explained (he did not specify what). Galileo also mentioned that he did not teach this new astronomy publicly for fear of the reaction from its numerous opponents.

This was the beginning of a long correspondence between Galileo and Kepler. Kepler correctly guessed that one of the things Galileo felt he could explain was the tides and he asked Galileo if he would make some astronomical observations on his behalf, believing Galileo to have access to more accurate instruments than he had.

One of these requests was that Galileo should try to observe stellar parallax, an argument frequently used to defend the old astronomy. The argument was quite simple. If the Earth were orbiting the Sun, our position in space would change and this change in position should make nearby stars appear to shift against more distant stars. This apparent change in position of a nearby object against a more distant background is familiar to anyone who has stared out of a car or train window as it whizzes through the countryside. Nearby trees appear to move backwards against more distant ones; this is parallax. Similarly, if you hold a finger up at arm's length and close one eye and then the other, your finger appears to move against the background.

No one had seen stars moving against more distant stars through the cycle of the year and this lack of stellar parallax seemed to imply that the Earth did not move. Galileo did not attempt the observation himself, having no hope of finding something that had eluded the best in the field. It was not until 1838 that stellar parallax was observed by German mathematician and astronomer Friedrich Bessel. He measured a parallax for the star 61 Cygni of 0.314 seconds of arc. (A truly tiny angle, as there are 3,600 seconds of arc in one degree of arc and 360 degrees in a full circle – 0.31 seconds of arc is the angle a US dime would make at a distance of thirty-three kilometres – twenty miles!)

It was while corresponding with Kepler that Galileo started a relationship with a Venetian woman called Marina Gamba and, although they never married, their first daughter Virginia – named after Galileo's elder sister – was born in 1600. In 1602 a second daughter by the name of Livia (after Galileo's younger sister) came along and in 1606 a son, named after his father Vincenzo, was born. Galileo's salary was good for a mathematician and he made extra cash on the side but he was still short of funds. He not only had to support his mistress and children, but also his mother and siblings.

The dowry of his sisters was meant to be split between Galileo and his brother but Michelangelo had no money. When Michelangelo got married in Poland it was Galileo who paid and later Michelangelo sent his family to live with Galileo, further draining finances. To make ends meet, Galileo had to take on extra private teaching, and borrowed money from a Venetian friend, Giovanni Francesco Sagredo.

Since arriving at Padua in 1592, Galileo had concentrated on refining his work on motion. He revisited his exploration of pendulums and the motion of objects down inclined planes. In late 1602 he wrote about this to del Monte and it is clear that Galileo was concerned with experimentally verifying his ideas. For example, Galileo realised his results with pendulums were more consistent if he used a heavy bob and long strings so that the angles through which the pendulum swung were small. It seemed paradoxical that, as the swing got smaller, the timing of each swing did not change. These observations probably led Galileo to the idea of inertia.

Inertia is fundamental to our understanding of motion and is at the heart of Newton's first law of motion (see page 48). This law can be stated as 'a body will either stay at rest or will carry on moving, unless it is acted upon by an external force', which is counter-intuitive. We know that if we give something a push it slows down and stops. But Galileo realised that motion would continue if it were not for friction and air resistance. He also recognised that inertia was why it was hard to get a stationary object moving in the first place.

By 1603 Galileo was devising experiments to study the acceleration of an object as it fell. Falling bodies were too swift to study properly but he realised that if he rolled balls down the inclined plane of a gentle slope, he could study the same effect in slow motion. Slopes of less than two degrees were still sufficient to see balls accelerate as they rolled. Until this time, although it was accepted that a falling object accelerated, it was assumed that this acceleration was made up of small successive bursts of speed, with the movement uniform between bursts. Galileo soon found that this was wrong.

By 1604, he devised a way to measure the acceleration of the balls as they rolled down the incline. Using musical beats separated by about half a second, he marked the position of the ball at intervals of equal time, and from these marks could measure the speed of the ball in each interval. He showed that the successive speeds in each time interval followed odd numbers, e.g. 1, 3, 5, 7, and that the total distance travelled went up by factors of 1, 4, 9, 16, etc. This gave the law that the distance travelled for a falling body is proportional to the square of the elapsed time.

These experiments highlight the break Galileo made with his contemporaries, replacing Aristotelian philosophical arguments with experimentation and measurement. The search for causes was now a search for physical laws – this is the key break that Galileo made with the past. Even astronomy, which had always involved making careful measurements, had left explanation to cosmology, which fell under Aristotelian rules.

In October 1604, while Galileo was in the middle of conducting these experiments, a star (that we now know was a supernova) appeared in the evening sky in the constellation Ophiuchus. A supernova is an exploding star but from Galileo's viewpoint it was a new, very bright star that he carefully observed. Aristotelian cosmology required the heavens to be unchanging, and so new stars had to be below the orbit of the Moon. If this were true, one would expect to see a parallax between the object and the fixed stars. But Galileo could not find any difference from observations made by various observers in different locations around Europe. He came to the conclusion that this new star was located in the celestial sphere, contradicting Aristotelian philosophy.

Galileo's dismissal of Aristotle soon brought him into conflict with Cesare Cremonini, professor of philosophy at Padua. Although they had been close personal friends, their dispute became very public, and escalated into a feud. Cremonini's arguments were put forward in a pamphlet published in Padua in early 1605, ostensibly by Antonio Lorenzini. Galileo recognised passages in it, which he felt had to be by Cremonini, and replied by publishing a response under an assumed name.

Cremonini suggested that the rules of measurement on Earth did not apply to the vast distances involved in the heavens. He also used the argument that the quintessence, from which the heavens were composed, was fundamentally different from the terrestrial elements, and this meant that the measurements could not be compared. Galileo was not convinced.

Over the next few years, Galileo refined his experiments on motion even further. He was able to establish that projectiles follow a parabolic path, something that proved crucial in military applications. He also noted that a ball rolling down a slope accelerated and one rolling up a slope decelerated, providing evidence for his inertia concept – that an object should neither accelerate nor decelerate on the flat. By 1609 he was hard at work composing a book on natural motions, though this was not finally published for many years.

At this point Galileo's career took an abrupt change of direction prompted by the arrival of the telescope. Galileo is sometimes said to have invented the telescope, but this is another myth. Hans Lippershey applied for a Dutch patent on such a device in 1608, which failed because telescopes already existed. Galileo's friend Paolo Sarpi heard about this invention before the end of 1608 and passed the details to Galileo on a visit to Venice in July 1609. Galileo immediately recognised a commercial application – that the telescope could be used by merchants to see ships approaching port earlier than with the naked eye.

Galileo hurried back to Padua to make the instrument that Sarpi had described. As he worked, he heard that a Dutch visitor had passed through Padua with a spyglass to sell to the Venetian government. Meanwhile, his friend Sarpi had been approached by the Venetians to ask whether they should invest in the Dutch invention. Sarpi advised against it, giving Galileo time to get to Venice in late August with an improved instrument, using one convex and one concave lens to give an upright image. He was able to show the Venetian dignitaries that they could see an approaching ship two hours before it would otherwise be visible. As reward, Galileo was offered lifetime tenure at Padua University and a near doubling of his salary.

Unfortunately, there were conditions. He had to remain on his previous salary until his contract finished and the new salary could not then rise. Finally, the offer stipulated that he remain at Padua for the rest of his career. Although Galileo liked Padua, he wanted to return to Florence at some point and intended to devote his time to research and writing, not teaching. He set about looking for another position, pursuing the job of court mathematician at Florence, where his friend Cosimo II de'Medici had become grand duke of Tuscany.

With this move in mind, Galileo arranged a visit to Florence to show his new instrument to Cosimo, then returned to Padua to work on a better one. He obtained glass blanks secretly so that none of his rivals would know what he was doing and ground the lenses himself. By early December he had built his third and most powerful tele-scope, with a magnification of twenty – nearly ten times the power of his first instrument. With this new telescope he set out to make his first astronomical observations and his target was the Moon.

Galileo could now clearly see that the surface of the Moon was not smooth, as taught by Aristotelian philosophers. The extra magnifica-tion showed detail invisible to the naked eye, including areas which should be in shadow near the Moon's terminator (the line dividing the lit from the unlit part) but were shining brightly. Galileo correctly interpreted this as being because the lit areas were higher than their surroundings and so caught sunlight when surrounding valleys were plunged into darkness.

On 7 January 1610 he turned his attention to Jupiter and noticed three star-like objects near the disk of the planet. He initially assumed that these were background stars but by 10 January he saw that one had disappeared and that by the 13th it reappeared, along with a fourth. For several weeks he continued his observations. These surely were not stars: they followed Jupiter across the sky, appearing to dance around the planet, sometimes with two on the right and two on the left, sometimes all four on one side.

Galileo realised that these points of light were, in fact, moons in orbit about Jupiter, just as our Moon orbits the Earth. He initially called them the Medicean stars, in honour of Cosimo. Now known as the Galilean satellites, all four are easily visible with a pair of binoculars. The closest, Io, takes under two days to orbit Jupiter and so, even over a few hours, Galileo could see a shift in its position. Discovering moons orbiting another planet was a huge blow to the Aristotelian system where everything was meant to orbit the Earth.

Galileo also turned his telescope to the Milky Way, the band of light stretching across the sky from horizon to horizon. He saw that this band was composed of thousands of individual stars, too faint to see with the naked eye. He hurriedly wrote up his findings in *Sidereus Nuncius* (starry messenger), published in Latin in March 1610 and also dedicated to Cosimo. This was the first publication of observations of the heavens made through a telescope. The public reacted with excitement but many philosophers and astronomers were dismissive, arguing that Galileo's observations were optical illusions.

In the summer of 1610, Galileo left Padua for Florence to live with his mother and take up the position as official mathematician at the Medici court. He had sent his daughters ahead, leaving his son with Marina as Vincenzo was only four years old. Soon after he moved to Florence, Galileo began observing the brightest planet, Venus. Earlier in the year it had been too close to the Sun to be visible, but now it had emerged from the Sun's glow and was accessible to his new instrument.

Venus is only visible in the morning or evening. The Greeks called morning and evening sightings 'Phosphorous' and 'Hesperus', the Romans 'Lucifer' and 'Vesper'. But as early as 1581 BC, a Babylonian

Figure 1, *opposite*: Galileo's telescope observations of the phases of Venus were crucial in showing that the Earth and Venus both orbited the Sun. Galileo saw that Venus went through all phases, from full to crescent; and appeared smaller when it was full and larger when it was crescent (top). These are naturally explained in a model which has both Venus and the Earth orbiting the Sun, with Venus' orbit lying closer to the Sun than Earth's orbit. In the Earth-centred model, Venus could only ever exhibit crescent phases (bottom).

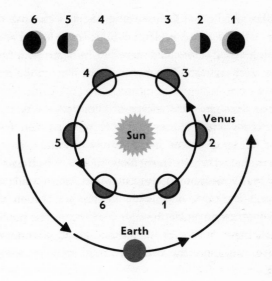

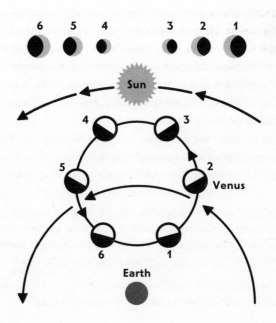

tablet makes it clear that the 'morning star' and 'evening star' were recognised as the same object. For months, Venus would be visible in the morning, rising before the Sun before disappearing into the glow of sunrise to re-emerge a few months later, rising and setting after the Sun.

Galileo's telescope observations of Venus proved that the Earth could not be the centre of the universe. It was clear that Venus went through a full set of phases, like the Moon. Having phases was not itself surprising – the geocentric model of the universe would have predicted that Venus show a crescent phase, with the side on which the crescent lay being dependent on whether the planet was a morning or an evening object.

Galileo, however, saw Venus go through all phases, not just a crescent. What's more, he could see quite easily that the planet appeared larger in his telescope when it was crescent and smaller when near full. There is no way these observations can be explained in the Aristotelian geocentric system but they are naturally explained in the Copernican heliocentric system (see figure 1).

If there was one piece of observational evidence that should have destroyed the geocentric model, this was it. But, possibly realising the controversy it would cause, the only person Galileo initially told was Kepler, in a letter in December 1610. He also mentioned that he had started work on measuring how long each of the four moons of Jupiter took to orbit the great planet. Kepler had his doubts that this could be done, but Galileo persevered and by March 1611 had enough data to predict the moments at which each of the moons would disappear behind the parent planet (called an eclipse of the moon).

Galileo was invited to present his findings at a meeting of the Lincean Academy, the world's oldest scientific society, founded in Rome in 1603. The banquet given for Galileo in 1611 is where the term 'telescope' was coined and the attendees were able to observe Galileo's discoveries for themselves using his instrument. Elected to the academy, Galileo found himself exposed to the academic debate and discussions on which he thrived. At the same time, he renewed his acquaintance with both Father Clavius and Cardinal Bellarmine. He gave frequent exhibitions of his telescopic discoveries in Rome, and these were well attended by prominent Romans, including several cardinals. Even Pope Paul V granted Galileo an audience.

Another astronomical phenomenon Galileo studied was sunspots. Although we now know that Chinese astronomers followed sunspots for many centuries, they were unknown in the West until the development of the telescope. Galileo discovered a book about sunspots by the German Jesuit Christopher Scheiner when visiting his printers. Assisted by a former pupil named Benedetto Castelli, Galileo took daily observations of sunspots and showed from their movement that they must be on the surface of the Sun, which he found rotated about once a month. Scheiner, by contrast, had concluded that sunspots were tiny planets orbiting the Sun very closely. Galileo published his work on sunspots in 1613 under the auspices of the Lincean Academy and they are known now as *Letters on Sunspots*.

In an appendix to the *Letters*, Galileo briefly discussed his work predicting the motions of the satellites of Jupiter. He knew that to predict these accurately it was necessary to introduce a correction for the Earth's changing position in space. He did not discuss this work in much detail, as he had realised that it might be possible to use the timings of the eclipses of Jupiter's moons as an accurate clock to determine longitude, a problem that had troubled surveyors and navigators for centuries.

Galileo was, by this time, convinced that he had observational evidence to prove that the Earth moved about the Sun. A deeply religious man, he was troubled that the Church was backing itself into a corner. Galileo wanted to separate scientific questions from matters of faith, and many of his actions in this critical period of 1613 to 1616 should be viewed in this light. He was not attempting to discredit the Church, but to stop the Church from backing the wrong horse.

Near the end of 1613, Galileo's old pupil Castelli, now teaching at Pisa, was invited to a court breakfast by Cosimo de'Medici. Also present were Cosimo's wife and his mother, the Grand Duchess Christina, with other members of the family. Another guest was a professor of philosophy at Padua, and when the subject of Galileo's observations of the moons of Jupiter came up, the professor pointed out to Christina that Galileo was wrong to argue that the Earth moved, as it was contrary to the Bible.

After breakfast, Christina took Castelli aside and asked him about the biblical miracle where Joshua stops the Sun in the sky. Castelli answered that scientific matters should be separated from teachings in the Bible. He sent an account of the incident to Galileo and Galileo wrote back to argue that freedom of enquiry should be allowed in all matters of nature which could be observed or about which experiments could be conducted. There could be no contradiction between nature and the Bible, but the Bible often spoke metaphorically.

Little happened for the best part of a year, but in December 1614 a Dominican named Thomas Caccini gave a sermon denouncing Galileo and the use of mathematics to understand nature. Caccini's sermon caused a stir. His motivation may have been to seek a more prestigious appointment in Rome but not everyone agreed with his tactics. Even his own brother wrote to him, urging him to stop attacking Galileo.

When news of Caccini's sermon reached Pisa, Nicolo Lorini, a colleague of Castelli's, expressed regret on seeing Galileo's letter to Castelli. Lorini took a copy to Florence, where he forwarded it to the Roman Inquisition. Galileo heard of this and fearing that Lorini had altered the letter before sending it on, obtained the original from Castelli and sent a copy to Piero Dini, a churchman in Rome, asking him to show it to the Jesuits and, if possible, to Cardinal Bellarmine. Lorini's partial copy had already been read at a meeting of the cardinals of the Inquisition, who asked the Archbishop of Pisa to get hold of the original and send it to Rome for them to study. A theologian wrote a report, finding that there were only a few phrases and words that were ill advised and that in general it was theologically unremarkable.

Galileo visited Rome towards the end of 1615, holding many public meetings to put the case for Copernican astronomy. In early 1616, he wrote *Discorso Sul Flusso e Il Reflusso del Mare* (discourse on the tides), which he sent to Alessandro, Cardinal Orsini. His theory depended on the Earth both rotating and orbiting the Sun, and Orsini took this discourse to Pope Paul V. The Pope instructed Orsini to tell Galileo that he must stop arguing that the Earth moved or the Pope would instruct the Inquisition to investigate him. Cardinal Bellarmine advised that the ideas be submitted to theological qualifiers. The Pope agreed and the qualifiers expressed the following opinions on Galileo's theories:

1. That the Sun is the centre of the world and totally immovable as to locomotion.
Censure: All say that the said proposition is foolish and absurd in philosophy and formally heretical inasmuch as it contradicts the express opinion of holy scriptures in many places, according to the words themselves and according to the common expositions and meanings of the Church fathers and doctors of theology.
2. That the Earth is neither in the centre of the world nor immovable but moves as a whole and in daily motion.
Censure: All say this proposition receives the same censure in philosophy and with regard to theological verity it is at least erroneous in the faith.

These censures were read out in the weekly meeting of the cardinals of the Inquisition on 24 February 1616. Pope Paul V asked Bellarmine to inform Galileo of the decision and to tell him that he could no longer hold or defend the ideas which had been censured.

If Galileo were to refuse to comply, the Commissary General would order that 'he must not hold, defend, or teach the propositions, lest the Inquisition proceed against him'. It was in interpreting this complex instruction that Galileo made his error. He assumed it was acceptable to teach the Copernican system as long as he gave both sides of the argument because that would mean he was not holding or defending the idea, as ordered by the Pope, and so the third stricture about not teaching the proposition would not come into force.

Meanwhile, Galileo was working on a refinement of his tables of the eclipses of Jupiter's moons, which he hoped could be used as a clock to determine longitude. He had already presented this idea to the Spanish government through their ambassador in Tuscany but it had not gone any further. By 1617 Galileo had achieved a remarkable level of accuracy in the eclipse times. Spain never took up his longitude scheme but it was adopted towards the end of his life by the Dutch government, which paid handsomely for the idea.

Galileo also returned to his treatise on motion. But in autumn 1618, just as he was immersing himself in this work, three comets appeared and Galileo's opinion was sought. Like new stars, Aristotelian philosophy required comets to be atmospheric, to occur in the sub-lunar realm. Galileo produced a long analysis, published in 1623 as *Il Saggiatore* (the assayer). It was here that he clearly outlined his ideas on scientific reasoning and how this approach could be contrasted to the untested arguments of natural philosophers. Galileo stated his belief in mathematics, saying that the book of nature was written in the language of mathematics.

The Lincean Academy agreed to publish *The Assayer*. Just as it was going to press, the pope died, replaced by Cardinal Maffeo Barberini as Urban VIII. The Academy dedicated the new book to Barberini as a fellow Florentine, an intellectual and an admirer of Galileo. In 1624, Galileo visited Rome to pay his respects to Urban, whose opinion seemed to be shifting from that of his predecessor. Galileo met the Pope on six occasions to persuade him to allow the publication of his theory on the tides. He pointed out that though his theory depended on the Earth moving, Italian leadership in science would be lost if the 1616 edict was enforced.

Galileo left Rome with the Pope's permission to publish his tidal theory as long as he made it clear that the Earth's motion was hypothetical. He was instructed to establish in the book that no experiments or observations could prove that the Earth moved. Galileo agreed and left Rome with the Pope's approval. Over the next six years, Galileo worked on *Dialogue on the Tides*. At the last moment, he was instructed to change its title, as the Church felt that it put too much emphasis on the Earth moving. Accordingly, Galileo renamed it *Dialogo Sopra i Due Massimi Sistemi del Mondo* (dialogue concerning the two chief world systems).

The dialogue was a popular way of presenting ideas to the public. In Galileo's *Dialogo*, the discussion is between two characters who try to convince a third – initially impartial – character. The pro-Copernican character is Salviati, named after Galileo's friend Filippo Salviati. The pro-Aristotelian is Simplicio, while the neutral character is Sagredo, after Giovanni Francesco Sagredo.

The *Dialogo* is a conversation over four days, during which Salviati and Simplicio discuss the relative merits of the Copernican and Aristotelian systems, trying to convince Sagredo. The first day involves a discussion about the difference in Aristotelian philosophy between the elemental make-up of the heavens and the Earth. The fundamental foundation comes in for criticism by Salviati, who states that new discoveries since Aristotle's time cast his ideas into doubt and that Aristotle made unproven assumptions.

The second day covers the Earth's rotation. Salviati proposes that the arguments used by Aristotelians against the Earth's rotation are flawed and uses Galileo's work on the conservation of motion and the relativity of motion to argue the Copernican case. Very little of Salviati's argument on this second day is astronomical.

On the third day, conversation turns to the motion of the Earth about the Sun. One of the main arguments used by Salviati is Galileo's work on sunspots. Salviati argues that this fits well with the Copernican model but is difficult to explain if the Earth is fixed and the Sun in orbit around it. Finally, on the fourth day, discussion concentrates on the tides. Although Galileo's tidal theory was wrong, he correctly points out that the tides are nigh on impossible to explain with a fixed Earth. Throughout *Dialogo*, Galileo tries to present both sides of the argument and to make the motion of the Earth appear no more than a hypothesis.

He had trouble obtaining a licence to publish *Dialogo* and, soon after he did, the Lincean Academy, which was sponsoring its publication, went into turmoil with the death of its patron, Prince Cesi. The book finally appeared in Florence in March 1632. It was published in Italian rather than Latin, showing Galileo's enthusiasm for it to be read widely by the public. But a few months after publication an order came from the Roman Inquisition to stop sales and Galileo was summoned to Rome to stand trial. Pope Urban VIII was furious, in significant part because Galileo had written the section of his book putting the official papal view in the voice of Simplicio.

When Galileo received news of his summons to Rome he was seriously ill and there was an outbreak of the plague, so his travel to Rome was delayed by several months. He finally arrived in February 1633.

The trial began on 12 April, the charge being 'vehement suspicion of heresy'. Galileo was asked the details of who handed the edict to him in 1616 and what he believed the edict to have said. Galileo replied that in February 1616 he had been told by Cardinal Bellarmine that since the opinion of Copernicus contradicted the holy scripture, it could not be held or defended. However, Galileo understood that this did not prevent him from discussing it using hypothetical arguments. But the Inquisition were only concerned about contravention of the edict and they found him guilty.

Before sentencing, Galileo acknowledged that he may have gone too far and that he would be willing to re-write offending sections if the Inquisition was lenient. He was, understandably, crushed when the sentence handed down was indefinite imprisonment. However, despite the myth, there is no evidence that he defiantly muttered '*Eppur si muove*' (and yet it moves) as sentence was passed. Galileo was not stupid.

After some pleading on his behalf, the Inquisition agreed to change the sentence to house arrest. Following a brief period at the Tuscan embassy, Galileo remained for the rest of his life at his villa in Florence. He lived eight more years before he died in 1642 at the age of seventy-seven.

In 1634 his *Mechanics* book, written about 1600, was translated into French, and the following year *Dialogo* was translated from Italian into Latin, enabling it to reach a wider audience in the academic community in Europe. Then, in 1636, his *Letter to Christina*, which had been circulating in manuscript form, was printed in both Italian and a Latin translation. From 1634 to 1637 he worked on his final and arguably greatest book, *Discorsi e Dimostrazioni Matematiche Intorno a Due Nuove Scienze* (discourses and mathematical demonstrations relating to two new sciences). Written again as a dialogue, bringing back the characters of the *Dialogo*, this book dealt with the structure of matter and the laws of motion. Two days of discussion were devoted to each of the topics and it brought to the public his long history of experimenting on motion. Because of restrictions placed upon him in Italy, the book was released for the first time by the publishers Elsevier in Leiden, Holland, in 1638.

This final book presaged much of Newton's work. Arguably, were it not for *Two New Sciences*, Newton may not have produced his masterpiece, *Principia* (though, interestingly, Galileo's book is not in the catalogue of Newton's library). Many of the important concepts of motion, such as momentum, relativity of motion, inertia, the acceleration of falling bodies and the parabolic motion of projectiles are outlined in *Two New Sciences*.

With the publication of *Two New Sciences* in 1638 and with blindness limiting his abilities, Galileo saw out the last years of his life in reclusiveness, dying peacefully in Florence on 8 January 1642, having altered our understanding of the universe and the way we do science forever.

At the end of the year of Galileo's death, a man was born who would start from his predecessor's position to scale new heights of understanding. That man was Isaac Newton.

....

CHAPTER TWO

Isaac Newton

·········

On Christmas Day 1642, the year that Galileo died, Isaac Newton was born. (At least, it was the same year in England. Italy had already adopted the modern Gregorian calendar and so, in Italy and in our modern calendar, Newton was born on 4 January 1643.) During a long career, Newton built on the work of Galileo to define a core physics that would last 250 years. Newton was the first to explain a force of nature – gravity – and the first to fully understand motion. His three laws bring the movement of objects to life and his analysis of light was masterful. It would take the genius of Einstein to show the limitations of Newton's thinking.

By the time Newton was born in Woolsthorpe Manor in Lincolnshire, his father, also Isaac, had died. As a baby, Newton lived with Hannah, his mother, in the Manor (little more than a large farmhouse). Although they were not rich, they lived comfortably. Their farm was tended by employees, while Hannah busied herself looking after her child and running the business.

This, however, was to change. Soon after Newton turned three, his mother married Barnabas Smith, rector of North Witham, a hamlet just over a mile (1.5 kilometres) from Woolsthorpe. Smith was considerably older, sixty-three to Hannah's thirty. His first wife had died some six months earlier and, as was typical of a widowed man of his time, he preferred to be looked after by a woman so had wasted no time in finding a new spouse. Hannah's brother, William

Ayscough, negotiated terms on her behalf. As part of this very businesslike deal, Hannah moved to live at North Witham, leaving Newton at the Manor with her parents.

The marriage lasted until Smith's death, eight years later, in 1653. During this time Smith and Hannah had three children, Mary, Benjamin and Hannah. There is little doubt that this enforced separation had a deep psychological impact on Newton. We know from the notebooks which he kept as a teenager that Newton hated his stepfather and resented his mother. Among a list of forty-five 'transgressions' from his childhood that Newton wrote when he was in his late teens there was 'Threatening my father and mother Smith to burn them and the house over them' and 'Wishing death and hoping it to some'.

When Smith died, Newton was eleven. Hannah returned to the Manor with her children from Smith in tow. The eight-year separation from her son had created a rift that never properly healed and Hannah may have been relieved that, a year later, Newton was old enough to be sent to King's School in Grantham, too far to travel daily. Newton boarded with local apothecary Mr Clark and his family above their shop on Grantham High Street.

King's was a prestigious establishment. Founded in the 1520s, it provided boys with a solid education in Latin, Greek and Bible studies. Newton initially found studies dull; he was far more interested in reading books from his stepfather's extensive collection than in his lessons. As a consequence, he was ignored by most staff and disliked by fellow pupils. One teacher did, however, encourage Newton in his reading, pointing out the book that ignited Newton's scientific mind, *The Mysteries of Nature and Art* by John Bate.

This book, published in 1634, was full of instructions for making tools and devices which Newton lapped up. He was able to design and build various working mechanical models for which he gained a reputation at school. Seventy years later, his biographer William Stukeley found people in Grantham who remembered the marvels

built by young Newton that included working windmills, kites, sundials and paper lanterns that Newton used to light his way to school on dark winter mornings. His talents made Newton more accepted by his fellow pupils, but he was never popular and remained indifferent to his lessons.

What appears to have transformed his attitude was a run-in with Clark's stepson, Arthur Storer. According to the story, one morning on the way to school, Arthur kicked Newton hard in the stomach. Upset, Newton challenged the heavyset Arthur to a fight. So enraged was Newton that he fought with a zeal that Arthur could not match and Newton humiliated him. Newton apparently told Arthur that he would beat him academically too and soon soared to first place in the school.

By all accounts Newton was otherwise happy with the Clarks and Mr Clark encouraged Newton to watch him at work. This experience was the first exposure Newton had to chemistry, leading to a lifelong obsession with – the then closely intertwined – subject of alchemy. Amazingly, he would devote more time to alchemy than to physics, although the extent of his work on alchemy only became known in the 1800s. During these formative years, he watched Clark at work and may also have experimented in concocting his own potions.

It was during this time that Newton developed one of his few close relationships with a woman. There remains considerable speculation about Newton's sexuality. Apart from his mother and his half-niece Catherine Barton, Mr Clark's stepdaughter Catherine Storer was the only female to whom Newton is known to have been close. Newton wrote to her during his early months at Cambridge and after his death Catherine would suggest that Newton had considered abandoning his studies to marry her, though this claim might have been an attempt to enjoy reflected glory.

As Newton's academic performance improved, he came to the attention of headmaster Henry Stokes. Stokes realised that Newton was bright enough to go to university but in late 1658, just as he was about to suggest this to Newton's mother, Hannah decided to remove Newton from school. In her view, reading and writing were sufficient for anyone.

Newton returned to live in the Manor, but it is clear from his notebooks that he was deeply unhappy. He lists some sins of 1659 as 'striking many', 'peevishness with my mother' and 'punching my sister'. He neglected farm duties to the extent of being fined in October 1659 for allowing his sheep to cause damage. Following this, Hannah decided that Newton's duties should be supervised by a servant but her son got the servant to do the work while he sat and read.

Under his mother's instruction, each Saturday Newton travelled to Grantham market with a servant to sell produce. Newton left the servant doing the work and visited Mr Clark where he sat in a back room, reading in solitary bliss. During these long hours, he gained a broader and more useful education than he had from the narrow school curriculum.

The servant complained to Hannah, while Mr Clark reported Newton's desire to continue learning to headmaster Stokes. Stokes decided to try Hannah again and she eventually relented, possibly because Stokes offered to waive the forty shillings (two pounds) charge for boys from outside Grantham. Hannah's brother, William – like Stokes, a Cambridge graduate – probably also helped persuade her. In autumn 1660, Newton returned to King's School to prepare for university.

He went up to Cambridge in the summer of 1661, when the study of 'natural philosophy' was in flux. Galileo had moved science towards the methods we know today, based on theory, experimentation and deduction. Newton was certainly influenced by Galileo, as well as by René Descartes, Robert Boyle and Francis Bacon.

In the 1660s, Cambridge was both academically backward and dangerous. The town's population of 8,000 included about 3,000 students, graduates and staff while the streets at night were haunted by prostitutes, beggars, thieves and murderers. Yet to Newton,

Cambridge was a paradise. When he left Woolsthorpe in early June 1661, he was embarking on the biggest adventure of his life. According to early Newton biographer William Stukeley, the servants at the manor cheered his departure and 'rejoiced at parting with him, declaring he was fit for nothing but the "Versity"'.

His mother ensured that Newton would not have it easy. Hannah paid Newton's fees but enrolled him at Trinity College as a 'subsizar', elevated a month later to 'sizar'. These were students who doubled as domestic servants for the more privileged, with chores that included emptying bedpans and cleaning rooms.

The structure of a Cambridge degree was very different in Newton's time. To obtain a Bachelor of Art (BA), a student had to reside at the university for a minimum of twelve terms (four years) and attend all public lectures by members of the college faculty. There was really only one course. The first year covered rhetoric – technically, the art of eloquent communication but including classical history, geography, art, biblical studies and literature. In addition, by the end of the first year, students were expected to be fluent in Latin, Greek and Hebrew.

During that year Newton was all but invisible. He was assigned an academic tutor, Benjamin Pulleyn, regius professor of Greek. Pulleyn was a 'pupil-monger', taking as many students as he could to supplement his income, leaving Newton one of over fifty undergraduates under Pulleyn's tutelage. Within weeks of arriving, Newton had cut himself off from his fellow students, making his first year at Cambridge lonely indeed. There is not a single anecdote from this time. All we know is that he appears to have detested his roommate, as he confessed to both using Francis Wilford's towel 'to spare my own' and giving him a bad name.

Newton's Puritan faith separated him from the largely Anglican university. This faith provided Newton with clear guidelines to cope with life but his pious intolerance hardly helped his popularity. And this was not all that alienated him. He decided to become a moneylender in his first year, and by the end of his second year, business was flourishing. He continued until he graduated and pursued his debtors with a zeal that resurfaced later in life when he became master of the Royal Mint.

During Newton's second year he shared rooms with John Wickins, son of the high master of Manchester Grammar School, who entered Cambridge in early 1663. The two lived together for over two decades, with Wickins later acting as Newton's assistant. They shared accommodation until Newton resigned as a professor and left Cambridge, but the nature of their relationship is not clear. When Wickins and Newton separated in 1683 they never met again. The feelings expressed in letters they exchanged have led some historians to speculate that they were lovers.

Throughout his first year, Newton attended lectures conscientiously but began to question Aristotle. In his second year in early 1663 he made a radical change. In the middle of a lecture, his notes stop mid-page. He then left dozens of pages blank before starting a new section headed *Quaestiones Quaedam Philosophicae* (some problems in philosophy). There he wrote '*Amicus Plato, amicus Aristotle magis amica veritas*' (I am a friend of Plato, I am a friend of Aristotle but truth is my greater friend).

Newton created forty-five query headings in his notebook, regarding the nature of the universe. Among these topics were 'Attraction magnetical', 'Of the Sun and planets and comets' and 'Of gravity and levity'. Under some headings nothing more is written, but others gained a paragraph or two of text and some are followed by a lengthy discussion. Newton initially copied the answers to these questions given by natural philosophers but then dissected the arguments. As an example, under 'Of gravity and levity' Newton wrote:

Try to discover whether the weight of a body may be altered by heat or cold, by dilation or condensation, beating, powdering, transferring to several places or several heights or placing a hot or heavy body over it or under it by magnetism, whether lead or its dust spread abroad, whether a plate flat-ways or edgeways is heaviest.

Newton was taking a new, scientific stance, questioning conventional wisdom. He sought out as wide a range of natural philosophy books as he could find in the library at Trinity College. Ironically, Galileo's two great books were banned by the college, but Newton may have read them in Grantham.

In summer 1664 Newton started experiments into the nature of light, prompted by buying a glass prism at Stourbridge fair. At the time, Descartes' was the most commonly accepted theory of light, arguing that it was a pressure, transmitted by the transparent ether to the optic nerve. Newton imagined light to be corpuscular, a stream of particles which he felt would make it easier to explain phenomena like reflection, refraction (the bending of light as it passes from one medium to another) and the distortions produced by lenses.

In a letter to Henry Oldenburg, secretary of the Royal Society, in 1672, Newton describes his earliest experiments with a prism:

> I procured me a triangular glass-prism, to try therewith the celebrated phenomena of colours. And in order thereto having darkened my chamber and made a small hole in my window-shuts, to let in a convenient quantity of the sun's light, I placed my prism at its entrance, that it might be thereby refracted to the opposite wall. It was at first a very pleasing divertissement, to view the vivid and intense colours produced thereby ...

With this prism, Newton was able to show that white light is composed of a range of colours, now called the spectrum of visible light. These colours (as defined by Newton) range from red to violet.

Newton also went on to correctly deduce why objects have different colours. The redness, yellowness etc. of an object, he wrote:

> are made in bodies by stopping the slowly moved rays without much hindering of the motion of the swifter rays and blue, green and purple by diminishing the motion of the swifter rays and not of the slower.

He was referring to red light as 'slowly moved rays' and blue as 'swifter rays'. We would say that a red object is red because it reflects the red part of white light but absorbs the other wavelengths, whereas a blue object reflects blue light and absorbs the rest.

Throughout the summer and autumn of 1664, Newton continued his experiments. He became so obsessed with his investigations that he took risks that could have left him blind. In one experiment he looked directly at the Sun, trying to observe coloured rings and spots before the eyes. In a letter to philosopher John Locke written more than twenty-five years later, he described how he was left unable to see for about three days and how he had to shut himself in his bedroom with the shutters closed to recover his eyesight.

A second experiment was even more foolhardy. To observe the effect of pressure on the back of his eye, he inserted a bodkin (a small dagger) between his eye and eye socket as far as the back of his eye and wiggled it around to see what effect it would have on his vision. He could easily have permanently blinded himself, ending his fledgling scientific career before it began.

It is also around this time that Newton started teaching himself mathematics; until then he had almost no grounding in the subject beyond basic arithmetic and geometry. By the end of summer 1664 he had mastered the most complex mathematical ideas of his time, from books like John Wallis' *Arithmetica Infinitorium* (1655), and Descartes' *Geometry*.

Newton nearly failed to pass an oral examination at the end of his third year that was required to continue into his final year. Although he had worked hard, it was almost entirely on material outside the curriculum. His tutor referred him to the first Lucasian professor of mathematics, Isaac Barrow, to be examined. Newton had skipped over most of Euclid's *Elements* but Barrow had just written a book on Euclid's work and was keen to explore Newton's Euclidian knowledge. Despite a faltering response, Barrow saw his potential and passed him. Having had his deficiencies illustrated so painfully, Newton ensured that he later knew the basics back to front. The most dog-eared book in his library was Barrow's *Euclidis Elementorum*.

Newton did little further preparation for his BA exams in spring 1665. At twenty-two, he graduated with an uninspiring second-class degree. By this time, the great plague had reached England and was killing thousands. Although Cambridge was less at risk than London, the university authorities sent all staff and students home as a precaution. Newton left Cambridge around the end of June, returning to Woolsthorpe for almost two years.

Legend has it that it was during this break that Newton made all his great discoveries, but this is an exaggeration. Certainly no apple fell on his head, though he himself claimed that it was *seeing* an apple fall at Woolsthorpe that inspired his thinking on gravity. His journals from this time show that his ideas on 'universal gravitation' were far from fully formed. Yet it certainly is true that Newton performed a remarkable amount of important work during his two years at home.

To take on universal gravitation, Newton had to develop a new form of mathematics. He was aware from Descartes' *Geometry* of the idea of fitting the gradient to a curve – effectively the angle of its slope at a point – something that Isaac Barrow had been researching when Newton was an undergraduate. Secluded in rural remoteness, Newton continued working on this problem, and came up with what we would now call 'differentiation'.

Differentiation allowed Newton to calculate the force felt by an object moving in a circle, such as a ball on the end of a string. He found that the force was related to the size of the orbit via the square of the radius. Newton extended this idea to the problem of the orbits of the planets about the Sun, and concluded that their endeavour 'to recede from the Sun will be reciprocally as the squares of their distances from the Sun.' On a scrap of parchment from this time we can see Newton's jottings of these calculations but it is only later that he outlined that this centrifugal force was balanced by a force of attraction (gravity) between the Sun and each planet.

By early 1667 Cambridge had reopened. Newton returned in March and started the struggle to secure a future by obtaining an MA and a fellowship. This process involved both the availability of a position and knowing the right people and it was crucial. If Newton could not obtain a fellowship, his academic career would be finished and in 1667 there were only nine positions to be filled from a total complement of sixty academics.

Newton devoted himself to preparing for the Trinity fellowship exams, held at the end of September. They consisted of three days of oral examination followed by a written paper. The results were announced at 8 a.m. on 1 October when a bell tolled to summon the candidates. Newton was successful and now had a job for life as well as the opportunity to study at his leisure. A small annual stipend of £2 and free rooms at the college were part of the privileges and when Newton obtained his Master's in spring 1668 his stipend increased to £2 3S. 4D. (nearly £2.70). Uncharacteristically, Newton even started socialising a little, visiting taverns, playing bowls and cards and relaxing his austere demeanour.

This did not last long. Newton soon returned to full-time work, developing his working relationship with Isaac Barrow. Barrow had entered Trinity in 1647 at the age of seventeen. There were no doubts that he was a brilliant mathematician, but he found himself constantly clashing with college and university. Vocal and political, Barrow was popular with his students but disliked by the authorities.

Soon after Newton was made a fellow, Barrow realised that his former student had made significant breakthroughs. He discovered that Newton was reluctant to publish and had to have anything he was working on teased out of him. Newton was suspicious that others would steal his ideas and kept his work largely secret, only revealing it to those who had gained his trust.

Barrow, on the other hand, was aware of the need to communicate in the scientific community and tried to persuade Newton to publish. An opportunity arose in September 1668 when Barrow was given a book on logarithms by Danish mathematician Nicholas Mercator and Barrow realised that Newton had gone far beyond anything in Mercator's book. Newton agreed to write a short paper summarising his work, entitled *De Analysi per Aequationes Infinitas* (on analysis by infinite equations), but then refused permission for publication. It did not appear until 1711.

When Barrow stepped down as Lucasian professor of mathematics in 1669 he recommended Newton as his successor and on 29 October Newton became the second holder of the post, aged just twenty-seven. In eight years he had gone from a freshman to a professor. He held this position until he resigned from the university in 1696, by which time he had become the most famous scientist in the world. This achievement was all the more astonishing because he accomplished it in his spare time from working on alchemy.

Alchemy had fascinated Newton since he read about it in Mr Clark's library. It may seem ironic that a chief architect of the scientific revolution should be interested in alchemy, but in a time before the atomic basis of chemistry was understood, alchemy and chemistry were impossible to separate.

Technically, 'operative' alchemy was illegal. A 1404 law outlawed making silver or gold and was not repealed until 1689. This meant that Newton conducted his studies in secret, although he kept extensive notes and, when he died, his library contained more alchemy books than those on physics (and more still on theology). His earliest alchemical notebooks date from 1669 although his writings on alchemy did not surface until the mid-1800s.

Newton was an obsessive and curious character and the mystical complexity of alchemy, together with his desire to unify and understand all of nature, was probably the reason he devoted so much time to it. It was also believed by many of its practitioners that alchemy would only succeed if the operator was 'pure of soul', which probably appealed to Newton's vanity.

Religiously, Newton held a heretical belief, known as Arianism, that rejected the Trinity and considered Christ to be the first creature. During the 1670s, Newton also studied obscure and ancient theological texts. This too became an obsession. Much of his theological work was published after his death in a book called *Observations upon the Prophecies of Daniel*, a rambling muddle of ideas.

Newton's first lecture as Lucasian professor, on optics, took place in January 1670. It was full of mathematics which the few students who attended probably found impossible to follow. For his second lecture no one showed up, and for the remainder of his time at Cambridge, Newton mostly lectured to an empty room. As his laboratory assistant described it:

> So few went to hear him and fewer that understood him, that ofttimes he did in a manner, for want of hearers, read to the walls.

This didn't matter to Newton. Teaching was an inconvenience and having no students was a bonus as he could cut lectures to fifteen minutes. His professorship required ten lectures a year but Newton rarely gave that many and soon compressed all his lectures into a single term. Throughout his twenty-seven years as a professor we only know of three students under his tutorage, Saint Leger Scroope, George Markham and William Sacheverell. None achieved anything academically and little is known about them.

Newton's first concern when he gained his professorship was to return to optics. In 1664 and 1665 he had shown that white light was composed of the colours of the rainbow and postulated that the colour of objects was caused by the way that they reflected light. When he returned to this work in 1669 he devised an experiment he called the '*experimentum crucis*' (critical experiment) that showed his theory of colour was correct.

His 1664 experiment, showing that white light could be split into its constituent colours, ran against the prevailing theory of light, which was that light was modified in passing through glass. Newton believed that the reason the white light was split was because the blue light was bent more in passing through the glass than was the red light.

To test this, he used a second prism receiving just the blue light from the first. He noted that when he passed blue light through the second prism he saw only blue light emerging; there was no further splitting of the light. Then, with the same experimental set-up, he allowed only red light to pass through the second prism. The red light emerging from the second prism was bent measurably less than the blue light, proving his theory.

Not content, he went a step further. Replacing the second prism with a lens, he focused the spectrum of light from the first prism and produced white light again. The lens recombined the spectrum of colours, reversing the work of the prism. Finally, he put a cogged wheel between the lens and the screen on which he could see white light, and, by carefully positioning the wheel, was able to block some of the colours from the lens before they reached the screen. When he did this, the spot on the screen was not white. (There has been some suggestion that Newton's equipment was not sensitive enough to produce the results he recorded, and he wrote down what he expected to see. Since he was correct, history has been kind.)

The first anyone in the scientific community knew of these experiments was in December 1671 and then only because Newton wanted to present his new reflecting telescope to the Royal Society. Unlike a refracting telescope, which uses lenses, a reflecting telescope uses mirrors. He had read about the Scottish astronomer James Gregory's failed attempt to construct a reflector and set about building a variant from scratch. His device was little more than 15 centimetres (6 inches) long but magnified about 40 times, better than a refractor of 1,8 metres (6 feet) in length. Barrow insisted that he show Newton's telescope to the Royal Society and reluctantly Newton acquiesced.

Barrow took the telescope, which was of exquisite design and craftsmanship, to London to demonstrate it to the fellows. Astronomer John Flamsteed, later the first Astronomer Royal, was captivated by its compact power. Christopher Wren (best known as the architect of St Paul's Cathedral but also a keen astronomer), Robert Moray and Sir Paul Neile took it to Whitehall to demonstrate to King Charles II. In early January 1672, Henry Oldenburg, secretary of the Royal Society, wrote to Newton saying the telescope 'had been examined here by some of the most eminent in optical science and practice and applauded by them'.

A few weeks later, Oldenburg again wrote to inform Newton that he had been elected a fellow of the Society. Newton was clearly delighted. In his reply he thanked the fellows, and proceeded to describe his theory of optics. By early February 1672 he had sent Oldenburg a detailed description of his theory of light, known as his *Theory of Light and Colours* letter. It was read out to the Society on 8 February and a few days after Oldenburg wrote to tell Newton it had been exceptionally well received.

Part of the Society's role was to reproduce experiments and this was the responsibility of Robert Hooke, curator of experiments, a position to which he was appointed in 1662. Hooke had attended Christ Church college, Oxford and was nominated for an MA in 1663. Hooke and Newton fell out within weeks, a situation that deteriorated to the extent that little more than twelve months later Newton was on the verge of resigning from the Society.

Unlike Newton, Hooke had broad interests. In 1665 he published his greatest work, *Micrographia*, which, in addition to covering microscopy and what could be seen under a microscope (it contains the first ever drawing of a plant cell – Hooke coined the term 'cell', thinking a row of cells looked like a line of the living spaces of monks), also contained a theory on light. Newton would have been familiar with this book and probably admired it. Yet their relationship became vitriolic and remained so until Hooke's death in 1703.

Newton wrote one of his most famous and misunderstood quotes in a letter to Hooke: 'If I have seen further it is by standing on ye shoulders of giants.' The phrase from February 1676 was probably quoted from Robert Burton's *Anatomy of Melancholy* and has been seen as self-deprecatory, implying that Newton's achievements depended on the work of others. But there seems little doubt that its main purpose was a putdown to Hooke, who was considered small and hunchbacked.

The conflict began because Hooke replied to Newton's letter on light and colour via Oldenburg, saying 'I confess I cannot yet see any undeniable argument to convince me of the certainty [of the theory].' Newton was apoplectic, but the custom of the time was to communicate politely. When Newton wrote, 'And having considered Mr Hooke's observations on my discourse, am glad that so acute an objector has said nothing that can enervate any part of it . . .' it is likely that he was seething with rage. To make matters worse, Hooke had found a genuine flaw in Newton's letter and took pleasure in pointing it out.

There was another correspondence that caused trouble between the two. Despite support for Newton's light theory from Huygens, one of the most celebrated optics experts in Europe, he was challenged by Jesuit priest Ignance Gaston Pardies. Newton wrote at length to Pardies, finally persuading him that the theory was correct. Hooke had been following this debate, undertaken through the Royal Society's *Philosophical Transactions*, and wrote to Oldenburg, complaining of Newton's conduct.

Oldenburg wrote to Newton in May 1672, suggesting that Newton tone down his language. After two weeks of silence, Newton responded, telling Oldenburg, 'Upon the receipt of your letter I deferred the sending of those things which I intended and have determined to send you alone a part of what I prepared.' He was referring to a new set of optical experiments and he refused to put them in the *Transactions*. As a consequence, this work would not be published until his book *Opticks* came out in 1704.

Before completely withdrawing, Newton answered Hooke's criticisms of his optics paper and these counter-arguments were read out at a meeting of the Royal Society in June 1672. In point after point, Newton demolished Hooke's objections in such a thorough manner that Hooke felt humiliated. This defeat opened a war that would rage between the two for the next 30 years.

Just when Newton thought he had silenced his critics, Huygens had a change of heart. When Newton received Huygens' objections in early March 1673 he wrote to Oldenburg:

Sir, I desire that you procure that I may be put out from being any longer a Fellow of the R. Society . . . For, though I honour that body, yet since I see I shall neither profit them, nor (by reason of this distance) can partake of the advantage of their assemblies, I desire to withdraw.

The issue of distance was an excuse – at that point Newton had not attended a single meeting – instead Newton's ego was bruised by criticism. For most of the first half of the 1670s he devoted himself to research into alchemy and the prophecies of the Old Testament.

By 1675, Newton felt emboldened to briefly re-establish contact with Oldenburg. The first of two papers, *An Hypothesis Explaining the Properties of Light*, explained how he felt reflection, refraction and diffusion of light were caused by corpuscles of light changing speed and being redirected as they travelled between media. His second paper, *Discourse of Observations*, discussed a series of experiments in which Newton attempted to demonstrate this theory, but this sparked a fresh battle with Robert Hooke who felt that Newton had stolen ideas from *Micrographia*.

Rather that state his objections through Oldenburg, Hooke held meetings in coffee houses where he vented his anger at Newton's 'theft', and news of these meetings found its way to Cambridge. A relationship that had begun badly got worse. Unusually for the times, the dispute between Newton and Hooke became direct. Hooke had fallen out with Oldenburg over several matters. Oldenburg was probably relieved not to be an intermediary and wrote to Newton warning him that Hooke would write to him directly.

This obliged Hooke to tone down his letters, as custom dictated that direct correspondence between gentlemen be civil. The correspondence began with a letter from Hooke in January 1676, replete with compliments to Newton and his work, although a deep dislike bubbles under the surface. Newton replied on 5 February 1676 and it is from this response that his quote about standing on the shoulders of giants comes. In subsequent private letters there is a regular subtext of mutual loathing.

The potential for a thaw between Newton and the Society came on 27 April 1676 – a day that is also a candidate for the beginning of the modern scientific age. This was the date of a Royal Society meeting at which the repetition of Newton's *'experimentum crucis'* was described, confirming Newton's hypotheses. But the relationship remained unchanged and in May 1677 Isaac Barrow died, followed by Oldenburg later that same year, leaving Newton feeling even more isolated.

Making matters worse, Oldenburg was succeeded as secretary by Hooke. Newton cut himself off entirely from the most important scientific society in England. He believed that he could not advance his ideas on nature if he was constantly answering questions from intellectual inferiors, so he locked himself away in a self-imposed isolation that would only be broken to spite Hooke.

During a coffee-house conversation in 1684 between Hooke, Christopher Wren and the astronomer Edmund Halley, the topic of planetary motions came up. Halley asked the others whether the force responsible for keeping the planets orbiting the Sun might decrease as the square of the distance. They told Halley that such a law was assumed. Halley asked if anyone had proven it to be true. Hooke claimed that he had done so some years before, but had kept his proof a secret as he wanted to see others fail before he showed how it was done.

Neither Wren nor Halley believed this bluster, and Wren offered a reward of a book worth 40 shillings (two pounds) if Hooke could produce his proof within two months. Hooke failed, and Wren stated, 'I do not yet find in that particular that he [Hooke] has been as good as his word.' Knowing that Newton was interested in gravity, Halley set off to Cambridge to talk to him. That decision changed the course of history. The meeting is recounted by a mutual friend, Abraham Demoivre, who said:

In 1684, Dr Halley came to visit him [Newton] at Cambridge, after they had been some time together, the doctor asked him what he thought the curve would be that would be described by the planets supposing the force of attraction towards the Sun be reciprocal to the square of their distance from it. Sir Isaac replied immediately that it would be an ellipsis. The doctor struck with joy and amazement and asked him how he knew it. Why said he, I have calculated it, whereupon Dr Halley asked him for his calculation. Without any further delay, Sir Isaac looked among his papers but could not find it, but he promised him to renew it and send it.

For three months, Halley waited patiently to hear from Newton, resisting the temptation to contact him. Finally he received a nine-page treatise, *De Motu Corporum in Gyrum* (on the motion of revolving bodies), but far more was to come. This document was the beginning of what would, two years later, be expanded into the *Principia*, arguably the most important physics book of all time.

The *Principia*, more properly *Philosophiæ Naturalis Principia Mathematica* (mathematical principles of natural philosophy), was published in July 1687 and is Newton's crowning glory, laying the foundations for 'classical' or 'Newtonian' mechanics. We still use the ideas in the *Principia* for almost all everyday applications of mechanics. Without it, Newton would still have been an important scientist, but the *Principia* elevates him to No 1.

The *Principia* is divided into three books. Book one, *De Motu Corporum* (on the motion of bodies), contains Newton's three laws of motion and his calculations on the elliptical orbits of the planets. Book two (uninspiringly titled *De Motu Corporum Liber Secundus* – on the motion of bodies, book two) deals with the motion of bodies through resisting mediums, including the effects of air resistance on a pendulum, and describes the way that the shapes of bodies affect their movement through resisting mediums. Book three, *De Mundi Systemate* (on the system of the world), contains his universal law of gravity.

As we saw in chapter one, Galileo devised the idea of inertia, which is formalised in Newton's first law, telling us that an object will stay at rest or will continue moving with a constant speed in a straight line unless an external force acts upon it. As Galileo realised, this is not obvious because objects do not carry on moving if we give them a push, because resistive forces slow them down. In the emptiness of space, Newton's first law is demonstrably true.

Newton's second law implies the crucial equation $F=ma$. In this equation, F is force, m mass and a acceleration. Most of the study of motion is built on this equation. From this equation, all the relationships that Galileo found experimentally can be derived.

Newton's third law, 'to every action there is an equal and opposite reaction', is the reason that you get a recoil from a rifle. The explosion which causes the bullet to come out of the front at high speed causes

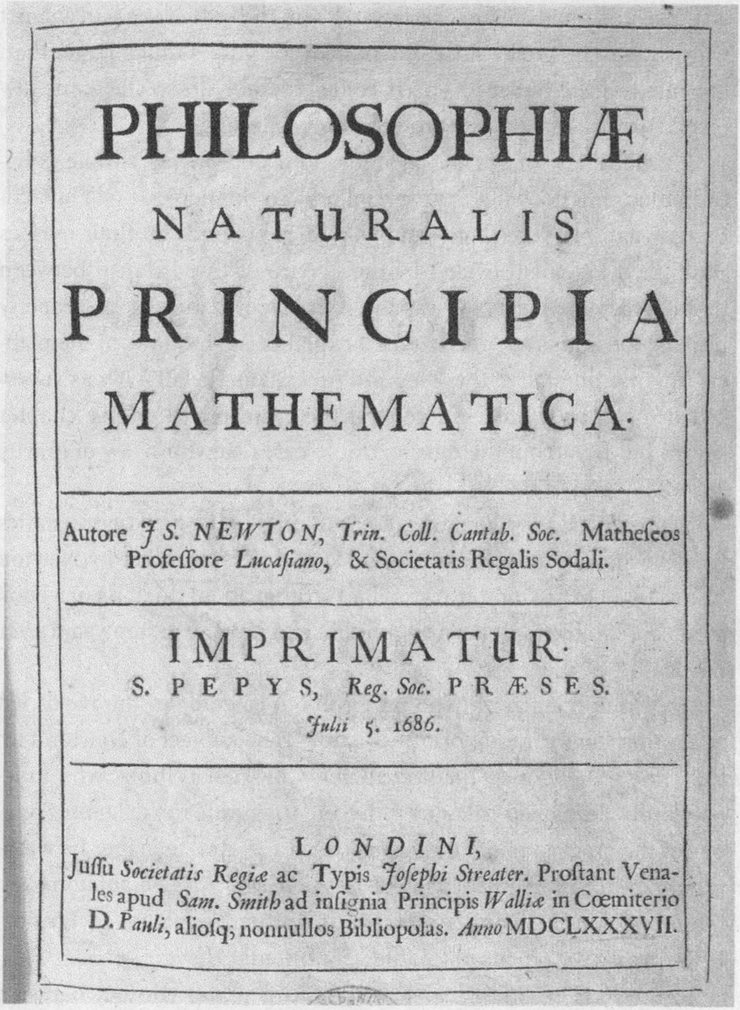

PHILOSOPHIÆ

NATURALIS

PRINCIPIA

MATHEMATICA·

Autore *J S. NEWTON*, *Trin. Coll. Cantab. Soc.* Matheseos
Professore *Lucasiano*, & Societatis Regalis Sodali.

IMPRIMATUR·

S. PEPYS, *Reg. Soc.* PRÆSES.

Julii 5. 1686.

LONDINI,

Jussu *Societatis Regiæ* ac Typis *Josephi* Streater. Prostant Vena-
les apud *Sam. Smith* ad insignia Principis *Walliæ* in Cœmiterio
D. *Pauli*, aliosq; nonnullos Bibliopolas. *Anno* MDCLXXXVII.

Figure 2: the front cover of Newton's *Principia*, arguably the most important
physics book ever written. The *Principia* was first published in 1687, and
Newton's own copy of the first edition (with his corrective notes in the margins)
can be seen at the Wren Library, Trinity College, Cambridge.

an equal and opposite force backwards – in the butt of the gun against your shoulder – but because the mass of the gun is much larger than the mass of the bullet its speed is considerably less. The same law enables a rocket to power through empty space.

Newton's law of gravity depends on a deceptively simple equation, though it does not appear explicitly in the book. It tells us that the gravitational force between two bodies depends on their masses multiplied together, divided by the square of the distance between them. This 'universal law' was the first law of nature to be properly understood and is so precise that it explains the motions of the planets, the rise and fall of the tides and how an apple falls. It was superseded in 1915 by Einstein's general theory of relativity (see chapter seven) but for all but the most extreme cases Newton's law of gravity suffices.

The *Principia* also contains the first hints of the new mathematics that Newton developed, now called 'calculus', but called by Newton 'the method of fluxions'. Newton had written about fluxions in a book completed in 1671 but it was not published until 1736, ten years after his death.

Newton's reluctance to go public about calculus in the 1670s led to another long-running priority dispute. The subject of Newton's ire this time was German mathematician Gottfried Leibniz, who independently developed calculus between 1673 and 1675. Leibniz was one of the greatest mathematicians of his age. As a teenager he wrote an important mathematics paper entitled *De Arte Combinatoria* (on the art of combination). As Huygens' influence waned in Europe, Leibniz rose to take his place, the 'Continental Newton'.

By 1675 it became clear to Oldenburg that a conflict between Newton and Leibniz was on the horizon. Oldenburg tried to convince Newton to publish his work on fluxions, but his appeals were met with silence. Leibniz published his first paper on calculus in October 1684, three years before the *Principia* and long before the wider mathematical community knew of Newton's mathematical work. In response, Newton added a section to book two of his manuscript:

> In letters which went between me and that most excellent geometer, G. W. Leibniz, ten years ago, when I signified that I was in the knowledge of a method of determining maxima and minima, of drawing tangents and the like and when I concealed it in transposed letters involving this sentence [he inserted the encrypted message he had sent to Leibniz in June 1676] . . . that most distinguished man wrote back that he had also fallen upon a method of the same kind and communicated his method, which hardly differed from mine, except in his forms and symbols.

The feud over calculus stretched until Leibniz's death in 1716. In 1712 the Royal Society commissioned a committee to investigate the matter and decide priority. By this time, Newton was president of the Society. He supervised every stage of the investigation and wrote the committee's inevitable conclusion that Newton was the first inventor. Although Leibniz tried to argue back – and now is considered an independent co-inventor of calculus – his health was failing and he died the following year, broken by Newton's relentless attacks.

When the *Principia* appeared, it proved dense and difficult to follow. Newton wrote it in Latin, which was normal for the time, so that it could be read by scientists across Europe. It was written in the classical manner, in the form of a series of propositions, each building on the one before. It was not a book to dip into idly (and remains so). Newton suppressed its publication in English until the final year of his life to reduce the possibility of lay people reading it.

The final book of the *Principia*, dealing with gravity, nearly didn't see the light of day. When Hooke heard a small section read out at a meeting of the Royal Society in April 1686, he was incensed to hear no mention of his own work on gravity. Halley conveyed Hooke's anger to Newton, who, in typical fashion, did not take it lying down. He went through the entire manuscript of the *Principia*, deleting any mention of Hooke he had made.

In a follow-up letter to Halley, Newton then wrote:

> The third [book] I now design to suppress. Philosophy is such an impertinently litigious lady that a man had as good be engaged in law suits as to she to do with her.

Halley had to use all his diplomacy to persuade Newton to send him the third part. He told Newton that he had consulted Christopher Wren, who agreed with Halley that Hooke's claims to any contributions to Newton's theory of gravity were outrageous:

> you ought to be considered the inventor and if in truth he [Hooke] knew before you, he ought not to blame any but himself . . .

Newton calmed down, and sent the third book to Halley in April of 1687, allowing the whole 550-page treatise to be published in July. It was the culmination of twenty years' work written in eighteen months. Newton had based his science on hypotheses, hard mathematical facts and verifiable evidence. Although the subject continued to be called 'natural philosophy', Newton's book was a masterpiece of physics.

With the publication of his book, Newton looked for new challenges. For his remaining years at Cambridge he became increasingly involved in university administration, and represented the university in the Convention Parliament convened to ratify William of Orange as King in 1689.

It is around this time that Newton met a young Swiss mathematician, Nicholas Fatio de Duillier. Fatio was twenty-two years younger, from a privileged family. After his arrival in England in 1687 he and Newton formed a close relationship. From the many letters exchanged between the two it seems possible that this relationship was sexual. Newton appeared infatuated with the younger man, showering him with gifts and affection.

In a letter Newton wrote to Fatio in November 1692, Newton signed off with 'Your most affectionate and faithful friend to serve you. Is. Newton' – the most demonstrative way Newton ended a letter to anyone. In early 1693, Newton suggested to Fatio that he move to Cambridge to be with him and Fatio replied:

> I could wish, sir, to live all my life, or the greatest part of it, with you, if it was possible and shall always be glad to any such methods to bring that to pass as shall not be chargeable to you and a burden to your estate and family.

In late May and early June of 1693, Newton made two trips to London to visit Fatio, but the second meeting marked the abrupt end of their relationship. It has been speculated that the reason for their splitting up was that Newton felt that Fatio was becoming too open about their shared enthusiasm for alchemy.

The impact on Newton was profound. By the end of the summer he appeared temporarily insane, sending rambling letters to the diarist Samuel Pepys and Locke. At this time Newton wrote a summary of his efforts in alchemy entitled *Praxis*. To quote biographer Michael White, '*Praxis* is little more than a blend of naked delirium and false conviction – the work of a man on the edge of madness.'

The break-up with Fatio was not the only factor that could have led to Newton's mental breakdown. He was lost, searching for a purpose. A lifeline came via his friend Charles Montagu, who got Newton a position at the Royal Mint. Newton left Cambridge for the last time on 20 April 1696 to take up his position at the Mint, based in the Tower of London. Initially he was warden, second in command to the master of the Mint. At the time, both jobs were sinecures.

The timing of Newton's arrival was fortuitous; the Mint was in a bad state of neglect. Many of the coins in circulation had been around since Elizabethan times, with some dating from Edward VI's reign 150 years before. As a consequence, many were worn out and the Mint faced counterfeiting and 'clipping', the practice of cutting bits off the edges of coins to collect the precious metals. Shortly before Newton arrived, a wide-reaching re-coinage was ordered to solve these problems.

The master of the Mint, Thomas Neale, drew his sizeable salary while expending minimal effort and the organisation was heading for a spectacular failure with Neale at the helm. But Newton threw himself into his new role, inspecting every aspect of the coining process. Newton turned up at 4 a.m. when the presses started and would often return to see the workers on the night shift, which ran from 2 p.m. to midnight. For his first few months he lived in the Tower, something no warden had done in generations.

Newton tracked down counterfeiters and clippers with a zeal that was reminiscent of his academic feuds. He conducted 200 interviews with witnesses, informers and suspects between June 1698 and December 1699, gathering much of the evidence himself. To make sure there was enough capacity for the re-coinage, Newton established branch mints in five other cities. When Neale died in 1699, Newton was promoted to master, remaining in post for the rest of his life.

Newton also mounted a successful campaign for the presidency of the Royal Society, the role that he had assumed by the time he oversaw the investigation into the claims of Leibniz. The previous president, Lord Somers, died in autumn 1703 and Newton was elected in November. By then, the Society had lost its way and its presidency had become popular with politicians with no interest in science. Newton's friend Montagu served as president from 1695 to

1698, attending only one council meeting. His successor, Lord Somers, was little better. The Society was losing money and authority but Newton would guide it back from the brink.

Soon after his appointment, in early 1704, Newton presented his new book *Opticks* to the Society. *Opticks* was first published in English and translated into Latin in 1706. It is in this book that Newton finally published his experiments from the 1660s and 1670s on the nature of light. *Opticks* was nearly as important to the study of light as the *Principia* was to motion and gravity. Newton had hoped to unify the macroscopic theories of motion that he had developed in the *Principia* with the microscopic 'corpuscular theory' he had developed for the nature of light, but had to abandon this effort, leaving it out of the published *Opticks*.

Newton's presidency consumed much of the last 30 years of his life. Almost as soon as he took over, he began an attempt to move the Society from its home at Gresham College but was initially hindered by dire finances. Three years later, Newton pushed through steps to charge admission and by 1710, the Society's finances were transformed. Newton suggested to the council in September 1710 that they move to a property he had found in Crane Court, London. During the move most of Hooke's equipment and the only portrait of him were 'lost'. Even with Hooke's death in 1703, Newton's vendetta continued.

Newton went on to have a third great feud, with John Flamsteed, the first Astronomer Royal at the Royal Greenwich Observatory. Flamsteed and Newton originally had a cordial relationship, when Newton asked for data to confirm his inverse square law of gravity. Their correspondence was brief, from December 1684 to January 1685, and Newton was extremely polite. In return, Flamsteed provided everything Newton asked for, including the positions of planets and stars, which allowed Newton to plot the path of the 1680 comet more accurately.

But as Newton took up the presidency of the Royal Society, their relationship soured. Newton had returned to his second edition of the *Principia*. Realising he was unlikely to get voluntary help from Flamsteed, who felt he had not been recognised for his contribution to the *Principia*, Newton decided to put Flamsteed under an obligation. Newton had gained royal connections and persuaded Prince George, Queen Anne's husband, to commission an astronomical catalogue from the Astronomer Royal, knowing that Flamsteed could not turn down the request.

Flamsteed planned a complete catalogue of the sky, *Historia Coelestis Britannica* (a British history of the heavens), showing the position of every visible star. He claimed that Prince George initially put up £1,200 to support the production of the work, over twice the cost of the Greenwich Observatory. But Newton persuaded George to reduce his investment to £863, forcing Flamsteed to narrow the scope of the work. Newton also arranged an expensive publisher minimising any financial reward Flamsteed got for his efforts. Understandably, Flamsteed felt bitter. Work moved at a slow pace, with Flamsteed objecting every step of the way, until October 1708 when Prince George died.

For a while the project was mothballed, but Newton persuaded Queen Anne to become patron of *Historia Coelestis* and in 1711 the project started again. Flamsteed continued to delay. At one point, Newton ordered Flamsteed to report his observations of the July 1711 solar eclipse and Flamsteed completely ignored him. Newton called Flamsteed before the Society's council to explain himself. Thirty years of frustration at Newton's hands came pouring out, creating, according to one Newton biographer, 'a public scene the likes of which had never before been witnessed at the Royal Society'.

Tired of Flamsteed's foot-dragging, Halley and Newton decided to use what data they had and published an unauthorised version of *Historia Coelestis* in 1712. Newton was then able to use the lunar data that Flamsteed provided to produce the second edition of the *Principia* in 1713. A furious Flamsteed referred to the publication as 'corrupted and spoiled' and called Halley a 'lazy and malicious thief'. Flamsteed would never see the publication of his own version, dying in 1719, six years before two of his friends ensured that it finally saw the light of day.

By this time Newton had moved to Cranbury Park near Winchester, owned by John Conduitt, whose wife was Newton's half-niece, Catherine Barton-Conduitt. Newton was only ever close to three women in the course of his life – his mother, Catherine Storer and now Catherine, who made him more sociable and interested in his family and also influenced the decor of his home. He split his time between this grand house and his house in Jermyn Street in London, where Catherine occasionally acted as hostess, until his death on 31 March 1727 (20 March 1726 back then, as years began on 25 March). He was buried, with a magnificent monument, in Westminster Abbey.

There are few similarities between Newton and the next physicist on our list, apart from both being British. Michael Faraday was as even-tempered as Newton was tempestuous, rarely holding a grudge, and rose to greatness without any formal training.

....

CHAPTER THREE

Michael Faraday

Michael Faraday is the only physicist in our top ten not to have received a university education but by virtue of his voracious appetite for self-education, his keen eye for experiment and meticulous attention to detail, he became one of the greatest experimentalists in history. In addition to important discoveries in chemistry, he discovered the principles behind the electric motor, the dynamo and the electric transformer and, by showing that light was affected by a magnetic field, he provided the first hints that light and magnetism were connected.

Faraday was born on 22 September 1791 in Newington Butts, about a mile south of Blackfriars Bridge. Today it is part of the London borough of Southwark but at the time it was suburban Surrey. He was the third of four children of James Faraday and Margaret Hastwell. The couple had moved to Newington Butts in 1786 from what is now Cumbria soon after being married. It was probably lack of work that caused James, a blacksmith, to move south.

James and his wife were Sandemanians, a small Christian sect that believed in a literal interpretation of the Bible, and, as a consequence, material wealth was shunned. Michael remained devoted to the movement's principles throughout his life. As a result, he refused a knighthood, twice refused the presidency of the Royal Society and would not be buried in Westminster Abbey. He refused invitations to the state funeral of the Duke of Wellington in 1852 and the marriage of Queen Victoria's eldest daughter in 1858.

When Faraday was four, the family moved to rooms over a coach house in Jacob's Mews near Manchester Square, on the western edge of London, now in the city of Westminster. James had started working for the ironmonger James Boyd in nearby Welbeck Street. The family was poor, and Faraday attended the local common day school. His basic education consisted of little more than learning to read, write and do simple arithmetic. Shortly after his thirteenth birthday, on 22 September 1804, Faraday went to work as errand boy for George Riebau, the owner of a bookshop and stationer's around the corner from the Faradays' home, in Blandford Street.

Clearly Faraday proved good at his job. Just after his fourteenth birthday, Riebau made him an apprentice to learn 'the art of bookbinding, stationery and bookselling'. Under the terms of the position, Faraday could not leave until the seven-year apprenticeship was finished. As a sign of his affection, Riebau agreed to waive Faraday's apprentice fee, saving Faraday's father a considerable sum of money. In addition, Riebau offered to house and feed Faraday in his home.

In October 1805, Faraday moved in with Riebau and his wife. By now he was taking a keen interest in the books that he was binding, particularly those pertaining to science. In 1809 he had started writing a journal, *Philosophical Miscellany*, making extensive notes on the books he read. These included scientific entries in the *Encyclopaedia Britannica*, *Conversations in Chemistry* by Jane Marcet and *Improvements of the Mind* by Isaac Watts.

The Watts book particularly influenced Faraday, teaching him skills of letter writing, how best to obtain knowledge and even how to keep a notebook. One of Faraday's greatest strengths would be the detail with which he noted down every aspect of his experiments, including unsuccessful ones. He developed an efficient information retrieval system, numbering every paragraph in his notebooks. This started in 1832 and twenty-eight years later his final entry would be paragraph 16,041. Faraday also started trying out the chemical experiments in Marcet's *Conversations*. Riebau allowed Faraday to set up a small laboratory with rudimentary apparatus.

The apprenticeship ended in 1812 when Faraday started work for a bookbinder, Henri De La Roche, at 5 King Street. Riebau helped Faraday secure the job, close to the family home, which had become important with his father's death in October 1810. Although Faraday had learned the trade, it is clear by the time that he started working for De La Roche that his heart was no longer in bookbinding. In a letter he complained that he wished 'to leave at the first convenient opportunity', as his true love had become science.

In the early 1800s there were few scientific posts. The only avenues to becoming a scientist (strictly a 'natural philosopher', as the term 'scientist' wasn't coined until 1834) were either to win a paid position (about a hundred of which were available in the UK) or to be independently wealthy. In 1812, one quarter of the Royal Society's 570 fellows had hereditary titles. For Faraday, the decision to leave a safe occupation must have required huge courage. His mother was a strong believer in providence and undoubtedly reassured Faraday that his life would follow God's plan.

While still apprenticed to Riebau, Faraday attended evening lectures by silversmith John Tatum at his house in nearby Dorset Street. Tatum advertised these events in handbills, open to anyone (men and women) upon paying one shilling (five pence) per lecture. Faraday later wrote that he attended 'twelve or thirteen lectures between 19 February 1810 and 26 September 1811'. Tatum's lectures were associated with the City Philosophical Society and enabled Faraday to meet like-minded people with whom he formed lifelong friendships.

Faraday took copious notes, writing them up, including diagrams, at his lodgings. He then bound them, completing four volumes of lectures. One day, in early 1812, Riebau showed the bound notes to a customer, the son of William Dance who lived in nearby Manchester Square. The following day, Dance senior came in to see the notes and was so impressed that he offered Faraday tickets to attend lectures by Humphry Davy, professor of chemistry at the Royal Institution.

The Royal Institution (RI) had been established in March 1799 by the Royal Society as an 'institution for diffusing the knowledge and facilitating the general introduction of useful mechanical inventions and improvements and for teaching, by courses of philosophical lectures and experiments, the application of science to the common purposes of life'. In today's language, the RI was intended to promote public engagement in science.

Fifty-eight of the Royal Society's fellows contributed 50 guineas (a guinea was twenty-one shillings or £1.05) each to become founding proprietors of the RI, with William Dance one of those founders. A building was bought at 21 Albermarle Street near Piccadilly in mid-1799, where the RI has been located ever since. The property, formerly a gentleman's town house, was refitted with laboratories and libraries. The crowning glory was a semi-circular, two-tiered lecture theatre that could hold more than a thousand people and is still in use today.

The first lecture delivered at the RI was on 11 March 1800 by chemist Thomas Garnett, but the Institution's star was Humphry Davy. He was appointed in February 1801 at the age of twenty-two and promoted the following year to professor of chemistry. Like Faraday, Davy came from humble beginnings. Born in 1778 in Penzance, Davy lived with his woodcutter father, attended the common day school in Penzance and then Truro Grammar School until 1795, when he started an apprenticeship as an apothecary.

He came to the attention of Davies Giddy, deputy lieutenant of Cornwall, a close friend of Thomas Beddoes, reader in chemistry at Oxford. In 1793 Beddoes fell out with the university authorities and left, setting up a tuberculosis clinic in Bristol called the Pneumatic Institute. Beddoes planned to treat the disease using gases recently discovered by Joseph Priestly and to do this he needed an assistant. Giddy recommended Davy, who moved to Bristol in October 1798.

Davy lived there for just under two years. It was in Bristol that he started experimenting with nitrous oxide, more commonly known as laughing gas, to which he became addicted. He published his account of it in his first book, *Researches, Chemical and Philosophical, Chiefly Concerning Nitrous Oxide*, in 1800. Davy moved to London in March 1801 to start working at the Royal Institution, which came increasingly under his influence.

Davy was a charismatic lecturer who sold out the theatre with his spectacular chemical demonstrations. The RI's income depended largely on ticket sales from its lectures so, once it became clear that Davy's demonstrations were popular, the Institution set about producing the best equipped laboratory in England. It was here, using an electric battery (invented by Italian Alessandro Volta in the 1790s) that Davy isolated for the first time several chemical elements, including sodium, potassium, calcium and barium.

Davy climbed rapidly in both career and society and was knighted in April 1812 by the Prince Regent, George III's son. Three days later he married a wealthy widow, Jane Apreece, and with her money, Davy was able to retire from RI duties at the age of thirty-four to pursue science as an independent gentleman. The Institution did not want to lose its best draw and so offered him the positions of Honorary Professor of Chemistry and Director of the Laboratory. As a consequence, his influence at the Royal Institution remained as strong as ever.

If the RI had hoped they would also retain Davy as a lecturer, they were to be disappointed. His lectures in March and April 1812 were his last. Faraday used Dance's tickets to attend each one, sitting in the gallery of the large theatre. The topic for this series was the nature of acidity, a hot topic in chemistry at the time. Through demonstrations, Davy showed, for example, that 'muriatic' acid (hydrochloric acid) did not contain oxygen as was then argued, but was composed of hydrogen and chlorine. Faraday took his usual detailed notes which he wrote up with diagrams and then bound.

In late December 1812, urged on by Riebau and Dance, Faraday wrote to Davy, enclosing his beautifully bound lecture notes. Faraday later wrote, 'The reply was immediate, kind, and favourable.' In Davy's response, penned on Christmas Eve, he said that he was delighted with the notes Faraday had sent him and with his 'great zeal, power of memory and attention'.

Davy said that he would like to see Faraday at the end of January. With what turned out to be sheer luck for Faraday, Davy had received an eye injury the previous month when an experiment combining nitrogen and chlorine exploded, driving glass fragments into his eye. He was obliged to employ someone for basic tasks and to take notes and it is clear that he had Faraday in mind.

Faraday's employment only lasted a few days while Davy's eyes recovered. But soon after, another stroke of luck came Faraday's way. On 19 February, William Harris, Superintendent at the RI, heard a 'great noise' in the lecture theatre. He found the laboratory assistant William Payne and John Newman, an instrument maker at the Institution, shouting at each other. Newman complained that Payne had hit him and, within a few days, the managers of the RI had sacked Payne. Davy set about finding a replacement for Payne and chose Faraday. He was twenty-two.

Technically Faraday was working for the RI, not Davy, but Davy monopolised much of Faraday's time. This became even clearer six months into the post, when Davy proposed to Faraday that he join him on a planned tour of Europe as his 'philosophical assistant'. Faraday had barely been outside of London, let alone beyond England's shores, but realised that being Davy's right-hand man was an opportunity he couldn't miss. On 13 October 1813, Davy, Lady Davy, her maid and Faraday left London.

They sailed from Plymouth for Paris on the first leg of a tour that would last eighteen months, travelling through France, Italy, Switzerland, Germany and Belgium. Faraday was expected to act as Davy's valet but in each city he also accompanied Davy to meet scientists to discuss the latest ideas in chemistry. When they reached Naples in March 1815, though, Davy learned of Napoleon's escape from Elba. With the political and military situation uncertain, the tour was cut short and Davy's party rapidly returned to England. For a few weeks after his return, Faraday was unemployed, but by the middle of May he was back at the RI. Faraday remained at the Institution for the rest of his working life, although he did not remain a lab assistant for long. In fact, Faraday was soon to help Davy make a major contribution to industrial safety.

Davy was asked by the rector of Bishopwearmouth to devise a way to safely illuminate coal mines. At the time, lamp-light underground caused horrendous explosions, as firedamp (methane gas) frequently exploded when exposed to naked flames. One of the worst disasters happened in May 1812 in Felling, near Newcastle in the north-east of England, with the loss of 92 lives.

With Faraday's assistance, Davy produced the Davy miners' safety lamp, which enclosed a flame in a fine metal gauze. The gauze allowed gases to pass through, but if firedamp ignited, the flame could not pass out and the combustion was contained. In addition, the condition of the flame served as an indicator of air quality. If there was too much carbon dioxide (which could cause asphyxiation) the flame extinguished; if the flame started burning vigorously, fire-damp was present. The Davy safety lamp saved thousands of lives in coal mines over the following decades.

In June 1821 Faraday married Sarah Barnard, daughter of a silversmith who was an elder in the church that Faraday's family attended. The families had known each other as long as Faraday could remember. She was nine years his junior – although in photographs from the 1850s she looks older than her husband. Sarah was one of three sisters and Faraday had long had a soft spot for her. They started exchanging letters but when Sarah's father saw one he intervened by sending her away to Ramsgate, a seaside town south-east of London.

Faraday could not endure the separation and hopped on a coach for Ramsgate to see her. He had been trying to summon up the courage to ask her to marry him for months. On a trip along the coast, Faraday took the bold step of holding Sarah's hand. That evening, he asked her to marry him and she accepted. They set the wedding day for 12 June. Faraday was living in a tiny room at the RI, so he approached Davy to ask if he could take over unused rooms in the attic. In May, permission was granted. Faraday was promoted to superintendent of the laboratory, although he stayed on the same salary of £100 a year, plus heat, candles and the attic rooms.

Not only was Faraday's personal life undergoing a massive change, but his scientific life also took an unexpected turn, inspired by the discovery of electromagnetism by Danish professor Hans Christian Ørsted. Ørsted had noticed that a compass needle was deflected from north when he switched the current flowing in a nearby wire on and off. He was able to show that a current flowing in a wire produces a circular magnetic field around the wire.

Ørsted's result received widespread attention. The French scientist André-Marie Ampère repeated the experiment and news reached the RI. Faraday did not pay much attention as he was busy with chemical experiments, but Davy did take an interest. In April 1821, William Wollaston, a medical doctor who dabbled in science, visited the RI and suggested to Davy that he and Davy perform Ørsted's experiment in reverse – to see if they could get a wire carrying a current to turn in a magnetic field.

Wollaston and Davy experimented for hours but with no success. Davy explained to Faraday what he and Wollaston had been trying to do, but Faraday paid little attention until asked to review the state of understanding of electromagnetism for *Annals of Philosophy*. Faraday wrote up the current position including the new result by Ørsted in an article entitled 'Historical Sketch of Electromagnetism'. But then he went a step further.

He cast his mind back to the unsuccessful experiment by Wollaston and Davy. Faraday realised that the wire had to be connected to the battery at both ends whilst still moving freely. In early September 1821 he devised an elegant but simple solution. He put one end of a bar magnet into the bottom of a basin and melted some wax which, once it had solidified, held the magnet upright in place in the middle of the basin. He next poured mercury – an excellent conductor of electricity – into the basin. One end of the wire stood in the mercury while the other was connected to a battery terminal.

Faraday connected the other terminal of the battery to the mercury, the circuit was completed and current flowed. As he did this, the wire started moving in a circle about the magnet. Faraday had discovered the principle behind the electric motor, calling the phenomenon 'electromagnetic rotations'. He wrote up his findings in the *Quarterly Journal of Science*, where they appeared as 'Note on New Electro-Magnetical Motions'.

The work was important in understanding electricity and magnetism, but it also marked the beginnings of a rift between Faraday and Davy. Wollaston suspected Faraday of stealing his idea. Davy supported his friend and reacted angrily to Faraday undertaking and publishing this work without consulting him.

The relationship between Faraday and Davy continued to worsen. Two years later, in 1823, Faraday liquefied chlorine gas for the first time. When Faraday wrote up his discovery, Davy felt that Faraday's account had not fully acknowledged his own contribution and made disparaging remarks about Faraday at a meetings of the Royal Society. In April 1823, Faraday's friend Richard Phillips, who himself had recently been elected a fellow of the Society, proposed that Faraday should also be given the honour. He organised the nomination but by the end of May, Faraday was told that Davy had asked Phillips to take down the nomination form.

Faraday, by this time much more confident in standing up to Davy, reminded his former mentor that only a proposer could take down his nomination. According to Faraday, 'Then he said, "I as president will take it down." I replied that I was sure Sir H. Davy would do what he thought was for the good of the Royal Society.' Davy did not carry out his threat, but spent time trying to persuade Faraday's proposers that he should not be elected. It was in vain; on 8 January 1824, Faraday became a fellow.

It might be that Davy's opposition to Faraday's election was in part because he could not support the election of someone with whom he had such a close working relationship. Davy was trying to move away from the patronage that was rampant during Joseph Banks' presidency. Whatever Davy's motives, Faraday later wrote that he was 'by no means in the same relation as to scientific communication with Sir Humphry Davy after I became a fellow of the Royal Society as before that period'.

Soon after, Davy involved Faraday in three time-consuming projects: the founding of the Athenaeum club, a project to protect the copper bottoms of naval vessels and an attempt to improve the quality of optical glass. The Athenaeum was an elite club for non-scientific people who otherwise might aspire to become fellows of the Society. For a month in early 1824, Faraday was tied up sending letters to eminent men, inviting them to join. At a meeting in May, Faraday was formally offered the position of secretary of the Athenaeum, with an annual salary of £100, but he declined. However, he did join the club and advised them on ventilation and lighting.

The previous year, Davy had been approached by the navy board to look into the corrosion of the copper bottoms of ships. Davy suspected that the corrosion was caused by an electrical reaction between the copper and oxygenated water and proposed that zinc could be applied to eliminate corrosion. The Admiralty ordered tests on three ships and in mid-February 1824 Davy's protectors were applied. Faraday undertook most of the follow-up experiments and by the end of April 1824 the navy board were satisfied that the protectors worked. The entire fleet was fitted out.

Unfortunately, early in 1825, problems became evident. Zinc protectors did reduce copper corrosion but left hulls encrusted in barnacles and other sea life by preventing the formation of the poisonous copper salts that had kept the creatures away. In July 1825 the Admiralty ordered the removal of all protectors and the blame was placed firmly at Davy's door. This public embarrassment damaged Davy's health and led to his resignation as Royal Society president in November 1827.

By far the most time-consuming of Davy's projects for Faraday was improving the quality of optical glass. A joint committee of the Royal Society and the board of longitude appointed glassmakers Pellatt and Green, with Faraday tasked to chemically analyse the glass produced. He later supervised the making of the glass, with the optician George Dollond grinding it and the astronomer William Herschel determining the glass's optical properties.

Davy's former patron Davies Giddy (now Davies Gilbert) replaced Davy as chairman of the joint committee after the zinc protector debacle and he decided to build a glass furnace in the RI for Faraday. High-quality optical glass has several requirements, including homogeneity and a high refractive index which is achieved by doping it with a heavy metal such as lead. As the glass cools the metal sinks to the bottom, requiring constant stirring. Unfortunately, this introduces bubbles and striations. A decade earlier, German scientist Joseph von Fraunhofer had overcome this but kept his process secret so Faraday's task was to replicate Fraunhofer's method blind.

In December 1827, Faraday attempted to reverse engineer Fraunhofer's techniques, a task that would end in failure. During the ensuing two years he spent two-thirds of his working time on the project, making 215 ingots of glass. Faraday became so deeply annoyed with the time the project was taking up that he started negotiating with the Royal Military Academy in Woolwich to become professor of chemistry. Then, in May 1829, Davy died a few months after suffering a massive stroke. The project was abandoned and Faraday became free to direct his own work.

One of Faraday's first tasks at the RI had been to help lecturers with demonstrations in the large lecture theatre. In the mornings, lectures were given for medical students studying at the nearby Windmill Street medical school, delivered by William Brande, who had replaced Davy as professor of chemistry. Afternoon lectures ranged widely with speakers including the engineer John Millington, the poet Thomas Campbell, the musician William Crotch, Peter Roget (of the thesaurus), and the architect John Soane.

On 7 December 1824, Faraday gave a lecture himself for the first time at the RI, although he had previously lectured between 1816 and 1818 at the City Philosophical Society. He began a 19-part series on metals for an audience that included the geologist Roderick Murchinson. He later remarked that Faraday's lecturing style was not good as he mainly read, failing to engage with his audience. With time Faraday would improve.

By December 1826 Faraday had become assistant superintendent of the house, a role that included overseeing the upkeep of the RI building. By now the most senior RI employee, one of Faraday's first acts was to address the perennial problem of revenue. He established two new series of lectures, both still running; the Friday evening discourses and the annual Christmas lectures adapted for a juvenile audience. From their inauguration in 1826 until his last in 1861, Faraday was the main lecturer of the Christmas series. These lectures have been televised for the last several decades.

While the Christmas lectures were open to all, the Friday evening discourses were only for RI members and their guests in formal dress. Faraday ran them from their inception until 1840, delivering about one in five – 127 lectures in all. He realised that the discourses could be used to increase membership of the RI and invited members of the press to help him in this. Lengthy reports of the lectures appeared in the popular periodicals and indeed the RI's membership did increase. In the early 1820s the RI was attracting about eleven new members a year; Faraday pushed this up to about sixty-five.

Clearly Faraday's lecture technique improved with practice. The RI started collecting data on audience numbers for the Friday discourses from 1830 and found that attendance at Faraday's lectures soared from about two hundred at the start of the 1830s to six hundred-plus by the beginning of the 1840s, levelling off at around eight hundred by the 1850s. On three occasions, over one thousand attended. Faraday became the most popular lecturer of his time.

Although Davy had kept Faraday busy, he had not stopped thinking about electricity and magnetism. Ørsted had shown that electricity flowing in a wire produces a magnetic field, but Faraday wondered whether a magnet could produce a current. He wrote in his notebook, 'If it is possible to convert electricity into magnetism, then why not the converse?' He made several attempts to achieve this in the mid-1820s, without success.

In August 1831 Faraday undertook a series of experiments to pin down the effect. He used an iron ring which he wound with two coils of wire placed on opposite sides of the ring. When he passed an electric current through the coil on the one side, he noticed that the needle on the galvanometer – a device for measuring electric current – that was connected to the other side briefly flickered, before returning to zero. It stayed at zero until he broke the circuit, when the needle flickered in the opposite direction. Faraday had discovered the principle behind the electric transformer that is today found in many devices – such as mobile phone chargers – and used extensively in our electricity distribution network in which high voltages from the power stations are reduced to lower voltages for domestic use.

Faraday was not entirely sure of the outcome, so he repeated the experiment and got the same results. He then took a break to visit Hastings with Sarah on holiday and did not get back to his laboratory until nearly a month later. Faraday continued to experiment with this new-found phenomenon and on 17 October he hit the jackpot, discovering how to generate an electric current by moving a permanent magnet in and out of a coil of wire.

Over several days he adapted this experiment, including placing a copper disk between the poles of a permanent magnet. On 28 October he found that when this disk was rotated an electric current was generated – a diagram from his laboratory notebook illustrating this is shown below.

Faraday had made possibly his most important discovery, how to generate electricity with a magnet and a conductor. He found that it did not matter whether the conductor was moved relative to the magnet or the magnet relative to the conductor; either motion generated an electric current. Every electrical generator, whether powered by coal, gas, nuclear or wind, uses this principle.

In March 1832 Faraday established another important concept when he found that the motion produced in a wire carrying a current in a magnetic field is at right angles to both. He also showed that the motion was greatest when the direction of the current and that of the magnet were at right angles to each other; despite his limited mathematical skills, Faraday was thinking in three dimensions.

Later in 1832 Faraday switched his attention to electrochemistry. At the time there was a dispute as to whether electricity produced by various methods such as by a battery, lightning, static, electric eels and Faraday's 'electromagnetic current' were identical or fundamentally different. Faraday believed that they were all the same, but was opposed by the RI's newly appointed professor of natural philosophy, William Ritchie, and by Humphry Davy's younger brother, John.

Faraday set about trying to establish that some observed phenomena such as the physiological effects, the magnetic deflection, the spark, the heating power and so on were the same, no matter what the source. To start his research, he submitted a paper on 15 December 1832 to the Royal Society in which he summarised the effects that had been shown to be common to all types of electricity.

When he reprinted the paper in 1839 he was able to fill in most of the boxes, based on his research in the intervening period, conclusively showing that the effects of each type of electric current were the same across a wide range of phenomena. With the help of William Whewell, a polymath who had introduced the terms 'Miocene' and 'Pliocene' for geological strata, Faraday introduced the terms 'electrode', 'anode', 'cathode' and 'ion'.

In trying to better understand the nature of electrical charge, in 1835 Faraday borrowed a large copper boiler, mapping the distribution and intensity of electricity inside of the boiler and on its surface. Copper is a good electrical conductor and relatively cheap, which is why we use it in most wiring. This was fine for an experiment, but Faraday wanted to demonstrate his results in the RI's lecture theatre. He constructed a large 'cube' of wire on a wooden frame, 3.7 metres (12 feet) along each side, placed on glass feet to electrically isolate it from the theatre's floor.

Faraday charged the wires surrounding the cube using static electricity and then, using a flap in one side of the cube, he stepped inside. In his own words, he 'went into the cube and lived in it', with 'lighted candles, electrometers and all other tests of electrical states'. He had showed that while there was an electrical charge on the

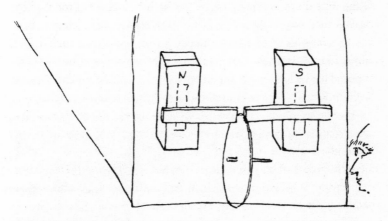

Figure 3: Faraday's drawing of the copper disk electric generator, as sketched by him in his laboratory notebook on 28 October 1831

outside of the cube, inside there was nothing. He had constructed what is now called a Faraday cage – the reason your car or an aeroplane can be hit by a bolt of lightning and you will not be fried by the huge voltage and current; no electrical charge penetrates inside the conductor. This illustration of a Faraday cage also helped him show that electricity was not a fluid, as some thought.

In 1836, Faraday was looking for possible connections between the known forces and wrote in his notebook that he would conduct experiments which would 'compare corpuscular forces in their amount, i.e. the forces of electricity, gravity, chemical affinity, cohesion, etc. and give if I can expressions of their equivalents in some shape or other.' However, despite noting down these thoughts in 1836, it was not until thirteen years later that he would turn his attention to addressing the question of the connection, if any, between these forces.

In June 1834 Charles Wheatstone had published a paper entitled *An Account of Some Experiments to Measure the Velocity of Electricity and the Duration of Electric Light.* Wheatstone had passed an electric current through a wire nearly 1 kilometre (half a mile) long, with spark gaps at each end and in the middle. He mounted a small mirror to the workings of a watch, so that the mirror rotated very quickly. He arranged his apparatus so that if the sparks were simultaneous, the reflections in the mirror would appear in a straight line. In fact, the middle spark lagged behind the others. The electricity had taken time to get to the middle of the wire. Wheatstone calculated the speed of electricity as 463,500 kilometres per second (288,000 miles per second).

This experiment piqued Faraday's interest. During 1836 and 1837 he varied Wheatstone's experiment, replacing one of the wires connecting the electrodes in the circuit with water, glass or another poor conductor. The time delay of the middle spark increased. Faraday theorised that there was a gradual accumulation of charge in these poor conductors that would discharge when large enough to produce a spark. In mid-November 1837 he introduced the idea of a 'dielectric' to explain the results, a dielectric being the electrical state of a non-conducting body between two conductors.

In the late 1830s Faraday turned his attention to the question of a connection between electricity and life. Some had argued that electricity was a life-force that could bring dead organisms to life. Ever since Luigi Galvani demonstrated that frogs' legs twitched when they were touched by an electrically charged scalpel, the idea that electricity played a central role in animation was popular – inspiring Mary Shelley to write *Frankenstein*. Faraday even kept some frogs in his laboratory to detect the presence of minute quantities of electricity.

Between 1839 and mid-1842 Faraday did relatively little research. In late 1839 he became ill with symptoms of vertigo, giddiness and headaches that forced him to retreat to the seaside for a month to recuperate. Throughout 1840 he struggled on but in December of 1840 the managers of the RI told him that he was relieved of his duties until he was completely recovered. He spent three months of 1841 in Switzerland, his first visit to the Continent since accompanying Davy. Faraday never fully regained his health and in letters he made frequent references to his poor health.

Despite his illness, Faraday was far from idle. In October 1840 he became an elder of his church, which entailed a great deal of extra duties, including leading worship. He also became heavily involved in government work, doing experiments for Trinity House, the Ordnance Office and the Home Office. Some have argued that he took on these government projects because he felt that his research had come to a dead end and did not know how to proceed.

If Faraday had lost his way, it was only briefly; by early 1843 he was speculating on the nature of empty space and if it could conduct electricity. In January 1844 he presented a lecture at the RI entitled 'A Speculation Touching Electric Conduction and the Nature of Matter', with an accompanying paper in *Philosophical Transactions*. He argued that the question of whether or not space conducted electricity could not be answered in terms of the atomic theory that the chemist John Dalton was proposing. Instead, Faraday argued that distributed throughout space were points where lines of force met. As part of this theory, Faraday realised that he would need to show that magnetism was a universal property of matter and not restricted to magnetic metals.

Faraday started a series of experiments in early 1845 to try to show this. He speculated that magnetism was a temperature-related phenomenon, as iron lost its magnetic properties when heated. Faraday wondered whether materials without magnetic properties at room temperature would become magnetic when cooled. By liquefying gases he was able to reach below −110°C (−166°F), and in May 1845 he started investigating materials at this temperature, seeing whether they became magnetic. Only cobalt did. Faraday published his results in the *Philosophical Magazine* and, despite his inability to get other metals to exhibit magnetism, he felt the only thing which distinguished iron, nickel and cobalt from other metals was the temperature at which they became magnetic.

In June 1845, Faraday attended the British Association's annual meeting where he met William Thomson, later Lord Kelvin, for the first time. Only twenty-one, Thomson was a rising star. The following year, Thomson would be appointed professor of natural philosophy at Glasgow University. At their first meeting Thomson and Faraday struck up a friendship which lasted the rest of Faraday's life.

In August, Thomson wrote to Faraday to ask him what effect a transparent dielectric would have on polarised light (light waves that only vibrate in one direction). Faraday's response was that he had tried that experiment in 1834 but found no result. Because of Thomson's letter, Faraday decided to reinvestigate the issue. As luck would have it, Faraday was testing four powerful lighthouse lamps for Trinity House and was able to use one to pass light through electrolytes (substances which ionise when dissolved) to look for effects, but again got negative results. He was then inspired to place the lead borate doped glass between the poles of an electromagnet.

This time Faraday found that the light transmitted through the glass changed its state of polarisation when the electromagnet was switched on. He wrote in his notebook 'magnetic force and light were proved to have relation to each other. This fact will most likely prove exceedingly fertile and of great value in the investigation of both conditions of natural force.' He called transparent materials that displayed the magneto-optical effect 'diamagnetics', in analogy with dielectrics. Faraday had made two important breakthroughs. Firstly, that light and magnetism were connected and secondly that glass could be affected by magnetism.

These results fitted Faraday's theory that all materials could exhibit magnetism under the right conditions. He next wanted to show that he could influence the glass directly with a magnetic force. He obtained half an anchor link from the shipping merchant Charles Enderby and used this to make a giant horseshoe electromagnet by wrapping about 160 metres (520 feet) of wire around the link. The final weight of his giant electromagnet was about 108 kilograms (238 pounds).

He started experiments on 3 November and on the following day hung a piece of 'heavy glass' (glass doped with lead) between the electromagnet's poles. He found that, when he turned the magnet on, the glass aligned between the poles. Within a week he had found that more than fifty substances could exhibit magnetic properties and in writing up these experiments Faraday first used the term 'magnetic field', introducing a concept which is central to our modern modelling of nature's four known forces.

By late 1851 Faraday had produced diagrams of iron filings around a magnet, an image with which every physics student is familiar. He first sent these diagrams in letters to friends and then in 1852 presented two papers where he used the diagrams as evidence for the existence of the 'field'. Faraday lacked the mathematical skills to put his theory into a mathematical framework; that work would be done by the next physicist in our top ten – James Clerk Maxwell.

Although the field concept began with Faraday's visualisation of magnetism, it was extended by Maxwell into the electromagnetic field and it was later extended to describe gravity and other forces. Even the famous Higgs boson has its own field, the Higgs field; through the Higgs boson, particles interact with the Higgs field and this gives them their mass. Einstein kept three portraits on his study wall: Faraday, Newton and Maxwell. In 1936 he wrote, 'The electric field theory of Faraday and Maxwell represents probably the most profound transformation experienced by the foundations of physics since Newton's time.'

In the late 1840s, Faraday met Prince Albert. The Prince had become vice-patron of the RI and in February 1849 made his first visit to a lecture by Faraday on diamagnetism. The two quickly developed a friendship and in 1855 the Prince brought his teenage sons, the Prince of Wales and Prince Alfred, to attend Faraday's Christmas lectures. In March 1858, when the new Chelsea bridge over the Thames was opened, Faraday accompanied Prince Albert and the Prince of Wales in the first group of people to cross the bridge.

The following month, a grace and favour house on Hampton Court Green became available which the Prince suggested to Queen Victoria would be ideal for Faraday. Later in the month, Faraday and Sarah visited the house, the Queen Anne building, but found it needed a great deal of work. Buckingham Palace undertook to pay for the repairs and in early September the house was handed over to Faraday. Despite keeping on their attic rooms, from that time onwards, Faraday and his wife spent increasing amounts of time at Hampton Court Green. It was there that Faraday died on 25 August 1867. He was buried in the non-Anglican section of Highgate Cemetery.

Faraday's experimental work on electricity and magnetism laid the foundations for the most important work of the next person on our list, James Clerk Maxwell. Whereas Faraday lacked any formal mathematical training, Maxwell was one of the most accomplished mathematical physicists of the nineteenth century and transformed the work of Faraday into one of the most important theories in physics – the theory of electromagnetism.

....

CHAPTER FOUR

James Clerk Maxwell

....

At No 5 in our list is Scottish physicist James Clerk Maxwell – far less well-known by the public than Galileo, Newton or Faraday, Maxwell deserves better. Einstein said, 'One scientific epoch ended and another began with James Clerk Maxwell.' The reason Maxwell is not more renowned is possibly connected to the highly mathematical nature of his work, although he was also a shy and retiring man who died young.

Maxwell was born in June 1831 in Edinburgh, the only son of John Clerk and his wife Frances Cay. Through marriage, the Clerk family had acquired the Middlebie estate, originally belonging to the Maxwells. Maxwell's grandfather bequeathed his baronetcy to Maxwell's uncle George, while the Glenlair estate, a 1,500-acre section of Middlebie, was passed on to the second heir, Maxwell's father John. When John Clerk inherited Glenlair he added Maxwell to his family's name.

John Clerk Maxwell was an advocate (barrister), making a good living from his practice in Edinburgh. He also had a private income from his family's wealth, but money was never a priority. When he met and married Frances Cay, the couple left Edinburgh for Glenlair, in the Vale of Urr, in Dumfries and Galloway, in south-west Scotland. John and his wife married late and their first child, Elizabeth, died in infancy. Frances was nearly forty when she had James; both parents doted on him. By now, John had all but given up advocacy, throwing himself into the life of a country gentleman.

Religion played an important part in James Clerk Maxwell's upbringing. Every morning the family and their servants met for prayer; every Sunday they trooped off to Parton church, eight kilometres (five miles) to the west of Glenlair. However, Maxwell's family were not puritanical. The Glenlair house was full of laughter and joy, with music and dancing a common occurrence. Maxwell was tutored at home by his mother from an early age but still played with local children, acquiring a thick Galloway accent that he never entirely lost.

The happiness of the Maxwell household was shattered when Maxwell's mother became ill with abdominal cancer, dying in 1839 aged forty-seven. John was too busy to tutor his eight-year-old son and felt there were no suitable local schools. Initially he employed a sixteen-year-old, who had delayed going to university, to tutor Maxwell. However, this experiment proved a disaster. The would-be tutor had no idea of how to stimulate the boy and Maxwell rebelled.

Thankfully, Maxwell's maternal aunt Jane offered help. She told John that she would look after Maxwell in Edinburgh with John's sister Isabella. A short walk from Isabella's house was the prestigious Edinburgh Academy; Maxwell could live with Isabella during term and return to Glenlair for the holidays.

Because the first-year class was full, the eleven-year-old Maxwell was put into the second year. Initially, things were difficult; not only was Maxwell surrounded by older boys but they made fun of his accent. Maxwell's clothes looked ridiculous to the sophisticated Edinburgh youths; they thought he looked and sounded like a peasant. Maxwell endured ridicule and bullying, often sitting alone in a corner of the playground. One classmate described him as 'a locomotive under full steam but with the wheels not gripping the track'. It would be more than a year before he made a real friend but this did not seem to bother him too much.

Maxwell exchanged letters with his father on a regular basis. Not surprisingly, these letters are full of childish jokes and stories, but just after his thirteenth birthday, Maxwell writes, 'I have made a tetrahedron and a dodecahedron and two more hedrons that I don't know the right names for.' He had not yet been taught geometry, but had read about the mathematics of 'regular polyhedra' and was making models of them.

The traditional rote learning style did not suit Maxwell but third-form work became more interesting and he was encouraged to apply himself. His results improved and he even grew to enjoy Latin and Greek. He now found himself surrounded by boys who were more keen to learn. When mathematics teaching started the next year his innate abilities seemed to be triggered. He astonished classmates with the ease with which he mastered geometry and, confidence boosted, began to do better in other subjects too. Soon he was in the top group of students.

About the same time, the class's star pupil, Lewis Campbell, moved into the house next door to Maxwell's aunt. Lewis was clever and well liked by fellow students. The pair became friends, often walking home together, discussing their schoolwork. Initially geometry was a common interest, but before long their discussions ranged far and wide, and their friendship continued until Maxwell's death.

Through Lewis, Maxwell became friends with other bright, lively boys in his year. One was Peter Guthrie Tait, who became one of the best physicists in Scotland, several times beating Maxwell to an academic position. Both became professors and continued to exchange letters throughout Maxwell's life.

Remarkably, Maxwell wrote his first scientific publication at fourteen, on curves drawn using string and pins. Tie one end of a piece of string to a pin and the other end to a pencil and you will draw a circle. Tie each end of the string to separate pins and push the pencil against the string to keep it taut – this produces an ellipse. The further apart the pins are compared to the length of the string, the more elongated the ellipse, while bringing the pins to the same place returns to a circle.

Maxwell learnt this in geometry but investigated further. He untied one end of the ellipse string and tied it to the pencil. He then allowed the string to loop from the first pin around the second pin and, keeping the string taut, drew a different curve, more like the outline of an egg. He went on to draw more curves, varying the number of times he looped the string about each pin. Maxwell derived a mathematical equation that related the number of times the string was looped around a pin to the distance between the pins and the length of the string.

Maxwell's father decided to show the work to James Forbes, a friend who was professor of natural philosophy at Edinburgh University. Forbes and his mathematics colleague Philip Kelland were captivated by the boy's ingenuity. They looked through the mathematical literature, discovering that René Descartes found similar results in the seventeenth century. Astonishingly, Maxwell's mathematics was simpler and more general than the Frenchman's. It later transpired that the equations Maxwell derived for his curves had a practical application in designing lenses.

Forbes presented Maxwell's paper to the Royal Society of Edinburgh, as Maxwell was too young to attend himself. The paper generated a lot of interest and Maxwell enjoyed the resultant attention. His father was thrilled with his son's achievement. But this playful work also marked an important step in Maxwell's scientific development. From this point on, Maxwell immersed himself in work by the great scientists; he also studied philosophy to better understand the scientific process.

During his teenage years Maxwell would sometimes stay with his aunt Jane, who ensured that Maxwell's religious upbringing was maintained. She took him to both the Episcopal and Presbyterian churches each Sunday, arranging for him to attend catechism classes, a form of Sunday school. Maxwell's deeply personal and reflective religious faith became the guiding principle of his life.

At sixteen, Maxwell entered Edinburgh University. He intended to follow his father into law but the curriculum was broad. He studied the classics, history, mathematics, logic, natural philosophy, 'mental' (classical) philosophy and literature. Although he found his first-year courses in mathematics and natural philosophy too basic, Maxwell was captivated by lectures in logic. Arguably, these classes developed two of Maxwell's most important abilities as a scientist. The first was his ability to return time and time again to the same problem, developing deeper insights. Secondly, because of this philosophical grounding, Maxwell was always comfortable with the notion that certain things cannot be measured directly. This enlightened view helped Maxwell develop the abstract equations of electromagnetism, and such thinking is fundamental today in modern physics.

Maxwell is usually thought of as a theoretical physicist. However, he had a passion for experiment kindled in his first year at Edinburgh by his father's friend, James Forbes. The two developed a rapport and Forbes allowed Maxwell to work alone late in the laboratory. Much of Maxwell's education during his three years at Edinburgh was acquired outside formal classes. During the six-month summer break that was the norm in Scottish universities at the time, Maxwell continued to conduct experiments in a makeshift laboratory at Glenlair.

One topic that fascinated him was polarised light. As we saw in the previous chapter, polarised light waves vibrate in one direction rather than in every direction – light reflected off a surface is polarised parallel to that surface, which is why Polaroid glasses reduce glare. Maxwell discovered that when polarised light was passed through unannealed glass (glass that was cooled too quickly, producing internal strains) beautiful coloured patterns appeared.

He extended this observation to see whether polarised light would show patterns in different solids under mechanical stress. He shone polarised light through clear jelly formed into doughnut-like shapes, twisting the jelly to stress it. The polarised light revealed internal strain patterns beautifully. This was the beginning of the photoelastic method still used by engineers to test designs. They build a scale model from a transparent material and view it with polarised light, subjecting the model to stress to observe the strain patterns.

Maxwell continued to write papers as an undergraduate and sent one initial draft to Forbes. The professor found the manuscript difficult to read, so poor was the structure. He gave Maxwell a rebuke, encouraging him to improve his writing style if he wanted his work to be read. In response, Maxwell developed one of the most fluent styles of any Victorian scientist.

Before Maxwell completed his degree, he persuaded his father that he could achieve more at Cambridge, applying to Peterhouse College, where his friend Tate had moved. Maxwell arrived at Peterhouse on 18 October 1850, travelling down with his father. During his first term, Maxwell's father discovered that there would be only one Peterhouse fellowship available to students graduating in Maxwell's year. They decided that Maxwell should switch to the larger and wealthier Trinity College.

Trinity was much more to Maxwell's liking. He was so keen to take advantage of all that Cambridge had to offer that sometimes he tried unusual routines. At one point he took to exercising in the middle of the night. A fellow student reported:

> From 2.00 to 2.30 a.m. he took exercise by running along the upper corridor, down the stairs, along the lower corridor, then up the stairs, and so on until the inhabitants of the rooms along his track got up and lay perdu behind their sporting doors to have shots at him with boots, hair brushes, etc., as he passed.

At the time Cambridge was primarily a training ground for judges and clergymen but even classics students had to tackle the Mathematical Tripos to get a degree. The first three days of the Tripos exams were standard book work mathematics, but, to gain an honours degree, a further four days of more difficult problems were set. These required insightful and creative problem solving.

Anyone who passed this seven-day ordeal, getting a first class honours degree, was known as a wrangler, something recognised throughout the world as an academic achievement. Wranglers were ranked by number and achieving senior (first) wrangler was akin to winning an Olympic gold. Straight out of the Tripos, the best mathematicians competed in an even more difficult exam to decide the winner of the Smith's Prize.

Maxwell graduated as second wrangler. The senior wrangler was Edward J. Routh of Peterhouse but, in the competition for the Smith's Prize, Maxwell upped his game and he and Routh were declared joint winners. Maxwell's immediate future was settled. He was able to stay as a 'bachelor scholar' at Trinity and could apply for a prestigious fellowship within a few years.

With the freedom to get on with his own research, Maxwell turned his mind to the process of vision – he was intrigued by how we perceive colour. There were no instruments to look into human eyes, so Maxwell 'invented' the world's first ophthalmoscope (he was unaware that it had already been invented by both Charles Babbage and Hermann von Helmholtz independently of each other). With this instrument he scanned the network of blood vessels that feed the retina.

Newton had shown that white light could be split into the colours of the rainbow. But how did humans perceive colour? Newton argued that colours such as brown that aren't in the spectrum could be made up by mixing other colours and created a colour wheel to explore how much of each of his seven colours were needed to create brown, pink, and so on.

Painters, on the other hand, mixed red, blue and yellow to produce their palettes. Textile makers mixed dyes similarly. Maybe there was something special about these three colours? English doctor and physicist Thomas Young suspected that the eye had three types of colour receptor for red, blue and yellow, but did not pursue the idea further.

Maxwell's interest in colour theory had been sparked in Forbes' laboratory. The professor had generated various colours by spinning a wheel made up of coloured sectors like a pie chart. Forbes found that no mixture of red, yellow and blue gave white. He took a step back and put just two segments on his spinning disk – one blue and one yellow – which produced green on an artist's palette. To his surprise, when he spun the wheel, it appeared dull pink.

This is where Maxwell picked up the research in 1854. He found a fundamental difference between mixing lights and mixing pigments. Pigments extract colour; the colour you see after mixing two pigments is whatever colours neither pigment has absorbed. Lights, by contrast add together to produce the final colour. Maxwell used this insight to construct a spinning disk of red, green and blue, which produced white. He constructed an apparatus where he would spin a disk with different combinations of those colours, comparing the result to coloured paper alongside.

Using a scale on the edge of the disk, Maxwell measured the proportion of each 'primary' colour (as we now call red, green and blue). From this experimental work he constructed what is now called the Maxwell colour triangle, the logic behind the production of colours on all modern TV and computer screens. He derived a mathematical formula giving the amount of each primary colour for any shade.

It was also during this period that Maxwell produced three papers on electromagnetism that heralded a revolution in the field. A lot was known about electricity and magnetism thanks to Faraday and his contemporaries (see chapter three), but this knowledge was fragmented with no coherent, overarching theory.

Until Maxwell's time, two approaches had been taken. Mathematical physicists derived equations that assumed electrical charges and magnetic poles acted at a distance, like gravity. Faraday, on the other hand, conceived of the idea that electrical charges and magnetic poles infuse space with a 'field'. In Faraday's model, lines of force emanated from electrical charges or magnetic poles and other charges and magnets felt these forces when the lines of force passed them.

Most scientists favoured the action at a distance that enabled precise formulae to be derived. These laws followed the familiar inverse square law that Newton had introduced for gravity. Faraday's idea of space being filled with invisible lines of force seemed crazy to most scientists. They doubted the merits of this self-taught man's ideas. Yet, reading Faraday's work, Maxwell came to recognise the power of the field concept. Faraday's notion of lines of force was a serious theory meriting further investigation. But it had no mathematical framework.

Stationary lines of force, Maxwell reasoned, were produced by a stationary magnet or a stationary electric charge. These lines of force would be three-dimensional and at every point in space a force acted with a particular strength along the line of force that passed through that point. The field was a collection of vectors, quantities that have both size (magnitude) and direction. The mathematics of vectors was in its infancy, but Maxwell was conversant with it and able to use it as he developed his framework.

Help came from work that William Thomson had undertaken on heat flow. Thomson discovered that the equations describing the strength and direction of the electrostatic force had the same mathematical form as the equations describing the flow of heat in a solid. Maxwell contacted Thomson and Thomson helped Maxwell come up to speed on the mathematics of what became known as vector fields.

Thomson drew an analogy between the lines of electrostatic force and heat flow but Maxwell envisioned electrostatic force differently. He imagined a weightless, incompressible fluid flowing through a porous medium. The lines of flow of this fluid represented the lines of force of either the electrostatic or magnetic field. By varying the porosity of his imaginary medium, Maxwell could account for materials with different electrical and magnetic characteristics.

In Faraday's model, lines of force were like tentacles, emanating from the magnet or electric charge. Maxwell modified this to a more continuous concept called 'flux'; the higher the density of flux, the stronger the electrical or magnetic force. The direction of the flow of the imaginary fluid at any point corresponded to the direction of the flux, and the speed of flow of the fluid was the 'flux density'. The fluid flowed due to pressure differences, as normal fluid flows in pipes.

Although Maxwell knew this was merely a model, it yielded the correct results. In addition, the model correctly explained some electrical and magnetic effects at the boundaries between different materials in a way that action at a distance could not. In Maxwell's model the fluid was incompressible, automatically producing the result that the electric and magnetic forces dropped off as the square of the distance.

Maxwell's next step was to see if he could use this model to write equations for the way that moving a magnet near a wire causes a current to flow. He took a single small part of space, re-expressing known laws in what we would call 'differential form'. This was not easy, but Maxwell discovered that the outcome was vectors that exactly fitted the idea that Faraday concocted to explain the flow of current when a magnet is moved near a wire. That was as far as Maxwell got in 1855. His equations described static electrical and magnetic effects and he knew he had to extend them to include fields that changed, but did not know how.

Maxwell wrote up the work in *On Faraday's Lines of Force*, presenting the paper to the Cambridge Philosophical Society. He emphasised that his model should not be taken as a representation of reality. He also sent a copy of the paper to Faraday himself, receiving the following reply:

I received your paper and thank you very much for it. I do not venture to thank you for what you have said about 'Lines of Force' because I know you have done it for the interests of philosophical truth; but you must suppose it is work grateful to me and gives me much encouragement to think on. I was at first almost frightened when I saw such mathematical force made to bear upon the subject, and then wondered to see that the subject stood it so well.

In February 1856 Maxwell received a distracting offer. Forbes mentioned a position as professor of natural philosophy at Marischal College in Aberdeen. Maxwell was only twenty-four, but several of his near-contemporaries were already professors. William Thomson had been appointed to the same post at Glasgow University aged twenty-two and Maxwell's Edinburgh Academy friend Tait became professor of mathematics at Queen's University, Belfast, at twenty-three.

After sending his application to Marischal, Maxwell spent the 1856 Easter holidays with his father who was recovering from a lung infection. John was clearly excited by his son's progress but tragically his illness got worse and John Clerk Maxwell died with Maxwell by his side. Maxwell was on the verge of a great academic career but also now responsible for the Glenlair estate.

Back in Cambridge, Maxwell received news that he had been selected for the Aberdeen professorship. At the end of the summer term, he packed up his belongings and left for Glenlair, spending the summer planning the running of the estate before heading for Aberdeen in November. He was the youngest professor in his new establishment by some fifteen years.

Marischal College provided a broad education to its students. Most hoped to go into careers in the law, the Church, teaching or medicine. In his inaugural lecture, Maxwell made it clear that his duties went beyond science to teaching his students how to think for themselves. In addition, he made it clear that experimental work would form a central part of the students' training, which was unusual for the time.

Maxwell took little interest in institutional politics but he could not fail to notice the chatter about the rivalry between Marischal and Aberdeen's other higher education institute, King's College. With only five universities in Scotland at the time (the others being

Edinburgh, Glasgow and St Andrews) many locals felt it ridiculous for Aberdeen to incur the expense of two establishments. A royal commission was set up to consider a merger and it was with this hanging over him that Maxwell started life in Aberdeen.

While at the college, Maxwell took on explaining Saturn's rings, which had puzzled astronomers for 200 years. St John's College Cambridge set the problem as the topic for its prestigious Adams Prize. In a series of elegant calculations, using existing mathematical techniques in new combinations, Maxwell showed that the rings could not be solid discs which would be unstable. But what about a fluid? Using the newly developed Fourier analysis, Maxwell showed that a fluid would break into blobs. By a process of elimination, he argued that the rings were collections of particles that were too small to see at our vast distance.

Maxwell submitted his work and was awarded the prize. His was the only entry but far from taking anything away from his achievement, this merely emphasised the difficulty of the problem. Astronomer Royal Sir George Airy pronounced Maxwell's work 'one of the most remarkable applications of mathematics to physics that I have ever seen'. In the 150 years since, no one has improved our understanding of the rings.

But Maxwell's most important work in Aberdeen was in the kinetic theory of gases. An understanding of the behaviour of gases due to heat had been developing during the latter years of the 1700s and the first part of the 1800s, leading to a new branch of physics called thermodynamics. Maxwell's interest had been sparked by an 1859 paper by German physicist Rudolf Clausius on the diffusion of gases. Diffusion explains, for example, how the smell of perfume spreads through the air.

Eighteenth-century Swiss mathematician Daniel Bernoulli had proposed a theory to explain diffusion that became known as kinetic theory. The idea was that gases consist of a huge number of molecules moving in all directions and the impact of these molecules on the surface of a container causes the pressure of a gas. Heat is related to the speed of the molecules. The higher the temperature, the faster the molecules travel and the more they batter the container. By the mid-1800s this theory successfully explained most observed characteristics of gases.

But diffusion presented a problem. At room temperature molecules had to be moving very quickly – several hundred metres per second. So why did it take so long for the smell of perfume to diffuse through a room? Clausius proposed that it was because the molecules kept colliding and were forever changing direction. To cross a room, a molecule might have to travel several kilometres.

Maxwell was intrigued. Clausius assumed that, at a given temperature, all molecules moved at the same speed. Maxwell felt that this was wrong and suspected that the speeds would be distributed about some average. To tackle the problem he could not use standard Newtonian dynamics because he was dealing with an assemblage of molecules with different speeds.

Maxwell took a statistical approach which had never been done before. He derived what is now called the Maxwell distribution for the speed of molecules in a gas. This distribution gives the range of speeds expected as a function of temperature. Maxwell had opened up an entirely new way of tackling physics. Had Maxwell done nothing else, he would still be one of the key physicists of the nineteenth century for this work alone.

During this time, Maxwell became a regular houseguest of the college principal, Reverend Daniel Dewar (unrelated to James Dewar, the inventor of the vacuum flask). Maxwell and Dewar's daughter, Katherine Mary, developed a close friendship over time. In February 1858 they became engaged and married in Aberdeen in June. Katherine was seven years older than Maxwell and they never had children but became lifelong companions.

In 1860, the royal commission recommended that there should be a single university in Aberdeen. As a result, there would only be one professorship in natural philosophy. Maxwell's rival, David Thomson, was sub-principal and secretary to King's College. The more senior and better-connected man got the job, leaving Maxwell unemployed.

He also failed to gain the chair at Edinburgh, where James Forbes was succeeded by his childhood friend Tait. But later in the year Maxwell won the professorship in natural philosophy at King's College, London. Over the summer, he wrote up his work on colour theory and sent it to the Royal Society in London. The society awarded him the Rumford Medal, its highest physics award. After a disastrous summer when Maxwell contracted smallpox and nearly died, he and Katherine set off in October 1860 for London.

King's College, on the Strand in central London, was founded in 1828 as an Anglican rival to non-sectarian University College. Unlike Cambridge or Aberdeen, King's courses had few relics of the medieval education system. There were classes in chemistry, physics, botany, geography and even engineering, courses not on offer in most other establishments.

Living for the first time in London, Maxwell was keen to attend lectures and discussions at the Royal Society and the Royal Institution. Although he had corresponded with Faraday for several years, they had never met. In late 1859, the two greats of early electromagnetism finally came together.

In May 1861 Maxwell was invited to give a lecture at the Royal Institution on colour theory. He decided to give a practical demonstration using the rapidly maturing technology of photography. Maxwell took three separate photographs of a tartan ribbon through red, green and blue filters. When combined, these separate images produced a colour picture – the world's first. (Luck was on Maxwell's side, as his photographic plate wasn't sensitive to red but did respond to ultraviolet which the red sections of the tartan happened to emit.) A few weeks after this demonstration, just before his thirtieth birthday, Maxwell was elected a fellow of the Royal Society.

After a five-year gap, Maxwell now returned to the connection between electricity and magnetism. He had always known that his fluid analogy would not work with fields that had changing values. Maxwell had to re-work the field in a new guise. He decided to change to a mechanical model.

The first problem of electromagnetism he tried to tackle was that opposite magnetic poles attract and like poles repel. The force is inversely proportional to the square of the distance between the poles and magnetic poles always occur in pairs. Maxwell searched out a medium that would fill space, producing magnetic attraction and repulsion. To do this he needed tension along the magnetic lines of force with pressure exerted at right angles to those lines. The stronger the field, the greater the tension and pressure.

The model Maxwell devised was a space-filling collection of tiny, close-packed spherical cells of low density that could rotate. When a cell rotated, centrifugal force would make it expand around its middle and contract along the spin axis, just as the Earth is thicker at the equator than the poles. As each spinning cell tried to expand, its neighbours would push back. Maxwell reasoned that if all the cells in a region spun in the same direction, each would push outwards against the others and collectively they would exert a pressure at right angles to their spin axes.

The opposite effect would occur along the axes of spin. In that direction the cells would try to contract and would set up tension. If these spin axes were aligned along Faraday's lines of force, the cells would produce an attraction along the axes of the cells and a repulsion at right angles to it. The faster the cells spun, the greater the attractive force along their axes of spin and the greater the repulsion at right angles – in other words, it would model a stronger magnetic field.

By convention, the direction of magnetic force goes from the north to south pole. To account for this, Maxwell added an additional convention that the direction of the field created by his cells depended on the direction in which they were spinning. Maxwell defined the normal direction as cells spinning in the same direction as a right-handed screw (the direction your fingers make if you curl them up in your right hand with your thumb pointing forward). Reverse the spin and the direction of the magnetic field is reversed.

But what set the cells spinning in the first place? And, if the cells jostled up against each other, spinning, say, clockwise, then the right-hand side of one cell which was moving down would rub against the left-hand side of the adjacent cell, moving up. This would create friction and stop the rotation.

Maxwell solved both problems with the same solution. He imagined even tinier spheres between the cells, like the ball bearings placed between wheel hubs to allow them to turn without friction. It seemed fanciful but it worked. These tiny spheres between spinning cells could be particles of electricity (electrons) that, when a battery was applied, would move along the channels between the cells and produce an electrical current. This motion of the spheres would set the cells spinning, producing a magnetic field, just as a magnetic field is produced around a wire carrying a current.

To account for the electrical conductivity of different materials, Maxwell suggested that this reflected the ease with which the tiny spheres could move. In copper, the spheres moved easily but in glass, cells or groups of cells held onto the spheres, stopping them moving.

A magnetic field without electric current made the spheres rotate with the cells. If the tiny spheres moved bodily without rotating they would 'drag' the cells on each side to spin in opposite directions, creating a circular magnetic field around a current-carrying wire. If the spheres moved and rotated, there would be a circular magnetic field superimposed on the one produced by the cells' rotation.

With this model, Maxwell could explain two of the four main properties of electricity and magnetism. But there was more to do – the next step was to explain how a changing magnetic field produces an electrical current. When a current is switched on in a circuit, it creates a pulse of current in a nearby wire – electromagnetic induction, discovered by Faraday (see page 71). Faraday had reasoned that the link between the two wires was made by a changing magnetic field created around the first wire when the current starts flowing. This changing field induced the flicker of current in the second wire.

Maxwell drew a series of diagrams to illustrate his model, showing the cells as hexagons rather than circles. A lower loop of wire included a battery, while an upper loop of wire did not. With a switch open (see Figure 4a) nothing moved. This was the situation with no current flowing, and no magnetic field.

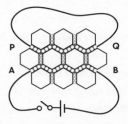

Figure 4a

In Figure 4b, the switch is closed, allowing a current to flow in the lower loop of wire. By convention, current flows from positive to negative. In Maxwell's model, the current is produced by the tiny spheres (or 'idle wheels' as he called them) physically moving, without rotating, from A to B, between the bottom row of cells and the middle row. This motion of the spheres caused the cells in the bottom row to turn in a clockwise direction and the cells in the middle row to rotate anti-clockwise. The two rows of cells rotated in opposite directions, as shown below:

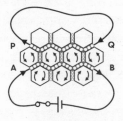

Figure 4b

The rotation of the two rows of cells created a circular magnetic field around the lower wire. Meanwhile, the spheres between the middle row of cells and the top row of cells were pinched between the middle row of cells rotating anti-clockwise and the top row of cells not rotating at all. This caused the spheres to rotate clockwise and to physically move from right to left, in the direction from Q to P, the opposite direction to the direction of motion of the tiny spheres in the lower part between A and B.

The loop containing the wire along PQ had resistance as all wires do, so the tiny spheres slowed down after they had received their initial surge of rotation, causing the top row of cells to start rotating anti-clockwise. The right to left motion of the tiny spheres soon stopped, but they continued to rotate. By this time, the top row of cells were rotating at the same rate as the middle row of cells (Figure 4c).

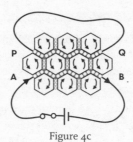

Figure 4c

The switch was now opened, causing the spheres between A and B to stop moving, so the bottom and middle rows of cells stop rotating. This means that the spheres between the middle and top rows (PQ) were pinched between stationary cells below and rotating ones above, starting them moving from left to right, the same direction as the original AB current (Figure 4d). Then, once again, the resistance of the upper loop of wire caused the spheres to slow down, but this time when they stopped moving they were also not rotating, returning to the situation in Figure 4a.

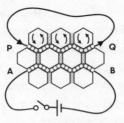

Figure 4d

To summarise, switching on a steady current in AB induced a temporary pulse of current in PQ, which moved in the opposite direction to the current in AB. When the current in AB was switched off, there was a pulse of current in PQ, in the same direction as the original AB current. Any change in the current in the AB circuit induced a current in the PQ circuit through the changing magnetic field that linked them. Here was a model which explained Faraday's discoveries.

Maxwell had explained three of the four observed effects of electricity and magnetism. So far he had not found a way to explain the electrostatic forces between electrical charges but his mechanical model explained the other three. Maxwell wrote his work up in *On Physical Lines of Force*, published in three instalments; part one in March 1861 and part two (in two separate instalments) in April and May 1861. He made sure to point out in his paper that his model was not a description of reality but explained observed phenomena using analogy.

In the summer of 1861 Maxwell and Katherine left London to spend some time at Glenlair. Maxwell planned to relax and work on the estate but his mind would not switch off. He realised that he had overlooked something important. For the cells to rotate together they had to transfer their rotational motion within the cell and to do this without losing any energy the cells had to be elastic. Maxwell wondered whether such elasticity could be the source of the electrostatic force.

In his model, current flowed in conductors because the tiny spheres were free to move in response to an electromotive force from a battery. In insulators, the flow of current was prevented because the tiny spheres were bound to neighbouring cells. But elastic cells could distort, allowing the tiny spheres to move a short distance. Because of the cells' elasticity, this distortion would produce a restoring force, encouraging the cells to spring back to their original shape. The tiny spheres would move until the spring-back force balanced the electromotive force supplied by the battery.

Maxwell reasoned that a battery connected by metal wires to two metal plates that were separated by an insulating material would produce a small displacement of the electrical particles in the insulator between the plates away from one of the conducting plates and towards the other plate. This small displacement, Maxwell realised, could be thought of as a brief electric current. But this small movement in the insulator between the plates would also be manifested in the wires connecting the battery to the plates and in the wires the tiny particles were free to move.

Therefore, the same brief current would flow in the wires too, leading to an excess of conducting particles on one of the plates and a deficit on the other. One of the plates would be positively charged and the other negatively charged and the distorted cells in the insulator between the plates would act like a wound-up spring, exerting a force of attraction between the plates. He was able to explain the (electrostatic) force between the plates by making the cells deformable and elastic.

When the battery was disconnected, the cells in the insulator remained wound up, storing energy. If the plates were then connected by a wire, this energy would be released as a current briefly flowing in the wire that connected the two plates. This would remove the charges on the plates; the cells and tiny spheres would also return to their rest positions in his model.

Maxwell reasoned that the degree of elasticity of his cells would depend on the nature of the material. In a good conductor, the elasticity would be small, represented by poor springiness. In insulators the elasticity would be large, leading to a greater electrical displacement for a given applied electromotive force.

He wrote up the detailed mathematics of this new addition to his model and it all hung together. He had shown that electrical and magnetic forces could be caused by energy stored in the space between and around objects. Electrostatic energy was a type of potential energy, like the energy in a coiled spring, while magnetic energy was rotational energy, like that of a flywheel; both could exist in empty space. Furthermore, the two types of energy were linked. A change in one always led to a change in the other and Maxwell's model was able to show how the two acted together to produce all known electromagnetic phenomena.

So far, so good. But, the model predicted two things that had never been observed. One was that there could be electric current anywhere, even in perfect electrical insulators and empty space. In a perfect insulator, there would be a twitch of current as the charges moved before their motion was stopped by the spring-back force of the elastic cells. Since all space was filled with the cells, these twitches of current would be present in empty space too.

Maxwell's earlier equations had dealt with the ordinary, familiar conduction current. But, now Maxwell realised that this new 'displacement current' had to be added to the equations. When he did so, his equations came together as a beautifully linked set. But his elastic cells predicted something even more interesting.

All elastic materials transmit waves. Maxwell's elastic cells filled all of space. So even in empty space, a twitch in one row of the spheres (the brief displacement current) would be transmitted to neighbouring cells and then to the next row of cells and so on. Because the cells had inertia, there would be a slight transmission delay between one row of cells twitching and the next. This meant that a change in the electric field would send a wave through space.

Furthermore, any twitch in a row of tiny spheres would make the neighbouring cells turn a little and a turning cell produced a magnetic field. Any change in the electric field would be accompanied by a change in the magnetic field and vice versa. The waves propagating through space would transmit changes in both the electric and magnetic fields. Because the changes were at right angles to the direction the wave travelled, Maxwell reasoned that such electromagnetic waves were 'transverse' – they wiggled from side to side.

It was already known that light was a transverse wave. Maxwell wondered if light was the wave his model predicted. The speed of light had been measured with increasing accuracy over the previous two hundred years and it was also known that the speed of waves in an elastic medium was given by the ratio of the medium's elasticity to the medium's density. In Maxwell's model, the elasticity of the cells was what controlled the spring-back forces, and the density of the cells determined the magnetic (rotational) forces.

When he plugged in the numbers he found that the speed of his waves in empty space or in air was exactly equal to the ratio of the electromagnetic and electrostatic units of electric charge (what we now call the 'permeability' and the 'permittivity' of free space).

He was able to calculate the speed of the waves in his model but had not taken any reference books to Glenlair so could not compare the result with the speed of light. Maxwell had to wait until he got back to London in October. His model predicted the speed of electromagnetic waves to be 310,740 kilometres per second (193,085 miles per second), and the latest figure for light, from an experiment by Frenchman Armand-Hippolyte-Louis Fizeau was 314,850 kilometres per second (195,639 miles per second). Surely these numbers were too close for coincidence? Light must consist of electromagnetic waves. Maxwell had made one of the most important discoveries in the history of science.

During the last months of 1861, Maxwell wrote up these results in parts three and four of his *Lines of Force* series of papers. Part three covered electrostatics, the displacement current and electromagnetic waves. Part four explained the magneto-optical phenomenon discovered by Faraday – that polarised light waves change their plane of vibration when they pass through a strong magnetic field. The papers were published in 1862. Maxwell was still not content, because he knew his model was not a true representation of nature. His next challenge was to make the model less artificial.

After publication, Maxwell took a break from electromagnetism. He instead turned his attention to the viscosity of gases, showing in a series of beautiful experiments that viscosity was constant over a wide range of pressures and that it did not vary with the square root of temperature, as was thought, but directly with temperature. This experiment diverged from the results of his model of the behaviour of gases and another problem was stacked in his mental processing pile to be returned to at a later date.

About this time, Maxwell was asked by the British Association for the Advancement of Science to bring some order to the chaotic system of scientific units. Magnetism, static electricity and current were clearly related, yet were measured totally differently. With help from Henry Fleeming Jenkin, Maxwell submitted a recommendation for a system of units that was adopted internationally and later known as the Gaussian system.

In 1865 Maxwell published the breakthrough paper *A Dynamical Theory of the Electromagnetic Field*. He abandoned his models and rebuilt the equations in his *On Lines of Force* papers from scratch. He turned to the work of eighteenth-century mathematician Joseph-Louis Lagrange, who had devised a way of deriving a set of differential equations describing the rate of motion of an object in terms of its momentum and its influence on the kinetic energy of the system. This approach treated the system as a 'black box'. If you knew the inputs and could specify the system's characteristics you could calculate the results without knowing how the system worked.

This was what Maxwell needed. As long as electromagnetism obeyed the laws of dynamics he could derive its equations without a model. It was a formidable task. Maxwell had to extend Lagrange's work to apply it to electromagnetism. He needed to keep certain key ideas in mind while developing his equations, such as the requirement for electromagnetic fields to hold energy which can do work even in empty space. Electric currents and the magnetic fields associated with them carried kinetic energy and electric fields held potential energy.

Although difficult mathematically, it came together remarkably well. Maxwell showed that the behaviour of electromagnetic systems, including the propagation of light, could be derived from the laws of dynamics. His work was stunning in its elegance and led to what we now call Maxwell's equations – four differential equations that link the six main quantities.

Maxwell published *A Dynamical Theory of the Electromagnetic Field* in 1865. He introduced it in December 1864 at a meeting of the Royal Society, but it is fair to say that nearly everyone who witnessed this talk was bemused. William Thomson never came close to fully understanding Maxwell's theory and he was far from alone. Most people simply didn't believe it. It was twenty years before German physicist Heinrich Hertz detected electromagnetic waves. Maxwell was asking his audience to believe in something that hadn't been shown to exist.

Meanwhile, Maxwell was frustrated by the pressures on his time at King's, despite being offered an assistant lecturer to help with his workload. He decided to resign his professorship and use his family's wealth to pursue his research independently. After five years in London he and Katherine returned to Glenlair in the spring of 1865.

Released from his duties, Maxwell undertook a flood of letter writing, particularly to Tait, still professor of natural philosophy at Edinburgh, and to Thomson. At the time, the pair were putting together *Treatise on Natural Philosophy*, a huge undertaking summarising the state of knowledge in physics. They asked Maxwell to check drafts of the manuscript. Maxwell himself was beginning to formulate his planned book *Treatise on Electricity and Magnetism*. This too would be a monumental task, taking seven years to complete.

In all, Maxwell spent six years at Glenlair. In addition to working on his *Treatise*, he published a book called *The Theory of Heat* and sixteen papers on diverse topics. He was also approached by Cambridge University to help revamp their Mathematical Tripos exam. The contents of this ordeal had not changed much since Maxwell took it, despite many advances in mathematics. Maxwell set about making the exam more interesting and relevant.

In 1866, Maxwell published *On the Dynamical Theory of Gases*, resolving a problem in his flawed paper of 1860. He and Katherine had conducted experiments on gases at home that showed the viscosity of gas did not vary as a function of the square root of its temperature, as his theory predicted. By assuming that the molecules repelled each other rather than colliding and acting like billiard balls and introducing the concept of relaxation time – the time a system takes to return to a state of equilibrium after it has been disturbed – Maxwell produced a resulting theory that the viscosity was directly proportional to the temperature.

Maxwell also wanted to test his prediction of the speed of electromagnetic waves. With Charles Hockin of Cambridge, he set up an experiment to measure the values of the magnetic permeability and the electric permittivity of space more precisely than had been done before. Their experiment used two metal plates, balancing the electrostatic attraction between the plates with the magnetic repulsion between two current-carrying coils. This required a high-voltage source and so John Peter Gassiot, a wine merchant in Clapham, was contacted as he had the biggest batteries in Britain in his private laboratory. He supplied 2,600 cells to Maxwell and Hockin, giving a total of about 3,000 volts.

Maxwell arranged for the experiment to take place during a visit to London in spring 1868. The batteries ran down so quickly that they became adept at taking readings quickly but found values for the permittivity and permeability making the speed of Maxwell's waves 288,000 kilometres per second (178,955 miles per second), within 3 per cent of the most recent measurement of the speed for light made in 1863 by French physicist Leon Foucault. Maxwell's theory that light was a manifestation of electromagnetic waves now stood on a more solid foundation.

Maxwell's *The Theory of Heat* introduced a new formulation of the relationships between pressure, volume, temperature and entropy in a gas. But the book also introduced a novel invention that took on a life of its own. He imagined a molecule-sized being who could intervene to make heat flow from a cold substance to a hot one, defying the second law of thermodynamics. Thomson called it 'Maxwell's demon', which soon caught on. In Maxwell's thought experiment, the demon stands guard over a hole in the wall separating two compartments filled with gas. The demon can open and close a shutter over this hole. Molecules in both compartments are moving in all directions and the average speed of the molecules is related to the temperature of the gas each side of the wall.

The thought experiment starts with the temperature equal on both sides of the wall but some molecules will be moving faster than average. The demon lets through faster molecules but blocks slower ones. Conversely, he lets the slower ones through from the other side and so changes the average temperature on one side to be hotter and that on the other side of the wall to be cooler.

This cannot happen, but the interesting question is why not? Maxwell's best explanation was that the demon would need to know the positions and the velocities of all the molecules, which seemed impossible. The solution was formalised in 1929 by Leó Szilárd, who showed that the act of acquiring information about a system increases its entropy (disorder), leaving less of the system available to do useful work. To gather enough information would counter any possible energy gains. Maxwell's demon helped spark the creation of modern-day information theory, crucial to both communications and computing.

When writing his *Treatise on Electricity and Magnetism*, Maxwell decided to simplify the equations using a system called 'quaternions', invented by Irish mathematician Sir William Rowan Hamilton. Quaternions are more complicated than vectors, having four components. Maxwell felt that this made his equations clearer, as in some cases he could replace nine symbols with two. He coined the terms 'curl', 'convergence' and 'gradient', now used by all students studying vector calculus (though they use the reverse of convergence: divergence). The job which Maxwell started was completed twenty years later by the American Josiah Willard Gibbs and Englishman Oliver Heaviside, who produced the modern notation for Maxwell's equations.

After a failed attempt to become the principle at St Andrews, Maxwell was asked in 1871 to accept a new professorship in experimental physics at Cambridge. The Duke of Devonshire, William Cavendish, chancellor of the university, had offered a large sum of money to his alma mater to build a laboratory. Cavendish had been second wrangler and Smith Prize winner in 1829. He was also great-nephew of Henry Cavendish, who first measured the value of 'G', the universal gravitational constant in Newton's equation of gravity, while determining the mass of the Earth.

Cambridge wanted a first Cavendish professor to start the new department and get the laboratory built. Maxwell was third choice after William Thomson and Hermann Helmholtz. He accepted the offer and was appointed in March 1871, leaving Glenlair with Katherine to return to Cambridge with a mixture of sadness and excitement.

In preparation for his new role, Maxwell visited the best university laboratories in Britain. He designed a laboratory that was so modern that it served Cambridge for the best part of a hundred years and was the birthplace of much modern physics. Lectures to new undergraduates began before the building was finished and experimental work began in spring 1874. Initially known as the

Devonshire building, Maxwell suggested it be changed to the Cavendish, commemorating both William and his great-uncle Henry.

In 1873, Maxwell finally published his *Treatise on Electricity and Magnetism*. A thousand pages of beautifully written text and mathematics contained everything known about electromagnetism. In it, Maxwell made an important new prediction – that electromagnetic waves exert pressure. He calculated that the Sun would produce 7 grams per hectare (0.006 pounds per acre), far too small to measure then but, twenty-five years later, Russian physicist Pyotr Lebedev confirmed Maxwell's prediction. Radiation pressure stops stars from collapsing under their own gravity as the outward radiation pressure balances the inward gravitational force. It also explains why comets' tails point away from the Sun, which had puzzled astronomers for centuries.

In 1868 and 1872 Ludwig Boltzmann had published papers on the kinetic theory of gases. Boltzmann derived a more general law for the energy distribution of molecules from Maxwell's work – now called the Maxwell-Boltzmann distribution. Maxwell was inspired to re-derive Boltzmann's more general results on a single page which he published in *Nature*. Even so, the theory that Maxwell had developed was still giving some wrong results. New observations showed that molecules vibrated as well as rotating. This produced differences between theory and observation that would require quantum theory to solve.

In the spring of 1877, Maxwell began to suffer from heartburn. He had increasing difficulty swallowing and in April 1879 mentioned this to the family doctor, who replaced meat in his diet with milk. As his attacks of pain became more pronounced, he was diagnosed with an advanced case of abdominal cancer. James Clerk Maxwell died on 5 November 1879 with Katherine by his side. He was forty-eight.

We encounter the products of Maxwell's work on electromagnetism and colour theory in modern technology every day. Maybe even more importantly, he started a revolution in the way physicists look at the world. He was the pioneer of the idea that maybe we only see a physical manifestation of something deeper, something we cannot directly perceive. Maxwell's equations of electromagnetism inspired Einstein to produce his special theory of relativity.

Our next figure was at the vanguard of unravelling the mysteries of further forces of nature – nuclear forces. Her name was Marie Curie.

....

CHAPTER FIVE

Marie Curie

At No 7 is Marie Curie, both the only woman on our list and the only one of our ten scientists to win two Nobel Prizes, in 1903 for physics and in 1911 for chemistry. The first woman to be appointed a professor at the Sorbonne, she discovered several elements, helped unravel the mysteries of radioactivity and dedicated her later career to using it to treat cancer.

Curie was born Maria Salomea Skłodowska on 7 November 1867 in Warsaw, Poland, the youngest of five – with three older sisters and a brother. Both her parents were teachers; her father, Władyslaw Skłodowski, taught science, and her mother, Bronisława, was a headmistress.

At the time Poland was carved up between Russia, Prussia and Austria-Hungary; Warsaw was under Russian control. Russian guards walked the streets and patrolled the schools. Classes were taught in Russian and teachers were not allowed to teach Polish history. Rather than extinguish the Polish spirit, this subjugation bolstered it. Curie's parents were among many hating the Russian regime. Władyslaw and Bronisława brought up their five children to defy and hate the Russians too. This fighting spirit helped Curie overcome obstacle after obstacle in later life.

She was a bright and curious child. Even before starting school, her father turned their chats into informal lessons. She was fascinated by her father's science equipment and enthralled by games he created to teach her and her siblings science, geography and history. Władyslaw also set his children mathematics problems;

Curie delighted in solving these problems faster than her older siblings.

When Curie started school, she was easily the brightest in the class and she had a phenomenal memory. Although classes were meant to be conducted in Russian, teachers used Polish when guards were not present. On one occasion a Russian guard paid Curie's class a surprise visit. Switching to Russian, the teacher asked star pupil Curie difficult questions which she answered with flying colours. She felt exhilarated but guilty for showing subservience to a Russian oppressor.

Life got harder for Curie's family over the years. Her mother was diagnosed with tuberculosis and had to quit her job, leaving the family dependent on her father's income. Soon after, her father was dismissed for disobeying his Russian superiors. To support his family, Władyslaw started a school at home; soon there were twenty boys attending classes, some boarding. The Skłodowska house was always crowded and full of activity, an environment in which the ever-curious Curie thrived.

Unfortunately, the home school was also a source of disease. Curie's sister Zosia caught typhus and died and a few years later her mother Bronisława lost her fight with tuberculosis. Carefree young Curie grew into an introverted teenager who seemed to carry the weight of the world on her shoulders. At the end of the academic year, her teacher recommended that Curie take a break to get over the loss of her mother but Władyslaw felt that a fresh challenge would help more. He sent her to a tough Russian school where Curie was awarded a gold medal as best student.

When Curie was fifteen, she was sent to live for a year with her uncle in the country. Away from the influence of her father, she was able to get up late, play outside like a child and indulge in many pursuits that had not been permitted, from fishing and collecting berries to playing games and going to dances. It was possibly the least stressful year of her life.

Back in Warsaw, Curie would have liked to move on to further education but, as was common then, Warsaw University did not admit women. Curie wanted to be a scientist like her father and decided to study on her own, but how could she learn science without access to laboratories? Luckily, a woman called Jadwiga Dawidowa had started up an unofficial university to teach the women of Poland.

Circumventing Russian rules, Dawidowa initially held classes in private homes before moving to bigger buildings, sometimes using school laboratories. Her institution had to keep changing location to avoid discovery. Dawidowa persuaded some of the best academics in Warsaw to give up free time to teach. Curie and her sister Bronia attended classes, though both knew that it was a temporary solution. Dawidowa's establishment could not provide recognised qualifications and so Curie and her sister looked elsewhere.

The best option would be to study at the Sorbonne in Paris; not only one of Europe's best universities but one that admitted women. The sisters agreed to take it in turns. Bronia would go to Paris first while Curie stayed behind to work, paying for Bronia's studies and saving for her own future.

Curie became a governess. Her first appointment was with a family which she later described as a 'family of lawyers ... where they don't pay their bills for six months and where they fling money out of the window even though they economise pettily on oil for lamps. They have five servants. They pose as liberals and in reality they are sunk in the darkest stupidity.' Curie hated the experience. 'I shouldn't like my worst enemy to live in such a hell,' she wrote to her cousin Henrietta Michałowska in December 1885. She left this family and took up a new position in the countryside outside Warsaw. This started in January 1886; just turned eighteen, Curie set off from Warsaw to the grand house of the Zorawski family, 80 kilometres (50 miles) north, in Szczuki.

The Zorawskis had three sons in university and living in the house were Bronka (eighteen), Andzia (ten), Stas (three) and

Maryshna, only six months old. Curie found the parents to be 'excellent people'. From the start she got on extremely well with Bronka (a shortening of Bronisława). However, being the same age made Curie's job tricky. She was Bronka's governess, yet they would have been classmates in school. Of the seven hours she was expected to teach each day, four were spent with Andzia, and three with Bronka. Curie delighted in playing with Stas. Her days were busy but she thrived on the atmosphere in the house. Curie also derived great pleasure from teaching a group of Polish peasant children to read and continued with her own education that she hoped would bring her closer to the Sorbonne.

In a letter to her sister, Curie waxed lyrical about a novel concerning romance between the classes called *On the Banks of the Nieman*:

> I've been shaken by reading a novel by Orzeszkowa, *On the Banks of the Nieman*. The book haunts me, I don't know any more what to make of myself. All our dreams are there, all the impassioned conversations which brought fire to our cheeks. I cried like a three-year-old. Why, why have these dreams faded?

Curie wrote her letter to Bronia in January 1888 in the middle of a personal crisis caused by her growing relationship with her employers' eldest son, Kazimierz. A year older than Curie, he was studying mathematics at Warsaw University. The pair made plans to marry but the Zorawskis refused a match between their son and a mere governess.

They did briefly rekindle their romance in the summer of 1887 but by December Curie had accepted that the marriage could never happen. For fifteen more months she worked with the family, initially in misery. By March 1889, though, Curie's mood had recovered.

Curie had still not heard from her sister so signed up as governess to the Fuchs family at a resort on the Baltic coast. After a year,

Curie finally returned to Warsaw, attending more classes at Dawidowa's university, knowing that her time to attend the Sorbonne was near.

Bronia wrote to Curie to tell her that she had finished her studies at the Sorbonne and had married. She invited Curie to Paris to live with her and to start her own studies. Bronia returned to Warsaw in the autumn of 1891 to help her younger sister prepare. Bronia had graduated in the July from the school of medicine, one of three women in a class of several thousand. She knew what was needed to succeed at the Sorbonne and was keen to pass on her knowledge.

So, in November 1891, nearly twenty-four, Curie boarded a train for Paris with a new first name, changing Maria to the French Marie. At the Sorbonne, she was in her element. She excelled at study and revelled in the freedom of her new environment. The only negative was living with Bronia and her husband – their cramped apartment doubled as a surgery and was crowded with patients all day long.

After six months, Curie found her own accommodation, closer to the Sorbonne, in the Latin quarter. She could only afford a tiny room on the top floor of an apartment building and so badly did she look after herself that, on one occasion, she fainted in the university library. In winter, the water in her washbowl froze overnight. Most nights she slept fully clothed with the contents of her wardrobe piled on the bed to try to keep warm.

When Curie moved to Paris she knew some French, but was far from fluent. She was one of only 210 women at the Sorbonne, out of over nine thousand students, and among twenty-three women in the 1,825 enrolled in the *faculté des sciences*. Initially, Curie socialised with a small colony of Polish students. However, by the end of her first year she gave up these friendships to concentrate on her studies.

When Curie arrived, the Sorbonne was undergoing a massive reconstruction. The Third Republic had made it the centrepiece of their educational overhaul of France. Science classrooms and laboratories were still being built, so Curie's classes were held in makeshift quarters nearby while architect Henri-Paul Nénot supervised the building of some of the most modern and best-equipped laboratories in the world. The science faculty doubled in size between 1876 and 1900. Money attracted talent. Curie would later write, 'The influence of the professors on the students is due to their own love of science and to their personal qualities much more than to their authority.'

Curie fully intended to return to Warsaw after completing her studies. She envisaged living at home until she met her future husband. But life took an unexpected turn. Graduating top of her class, she was offered a scholarship to stay at the Sorbonne, studying for a *licence ès mathématiques* (the equivalent of a BSc in mathematics). She could not turn this down; science was her life.

In the winter of 1893–94 Curie was also preoccupied with what would become a perennial problem – the search for more laboratory space. She was well along in her preparation for her degree in mathematics and had been hired to study the magnetic properties of various steels. She conducted this research in the laboratory of her professor, Gabriel Lippmann, but it was cramped and poorly equipped.

When some Polish friends, Józef Kowalski and his wife, visited her in spring 1894, she moaned about these conditions. Kowalski, a physics professor at the University of Fribourg, knew of someone doing similar research nearby, a Frenchman named Pierre Curie. Pierre had become famous at only twenty-one when he and his brother Jacques discovered that quartz crystals could hold an electrical charge. After that he invented the electrometer, which became the instrument of choice to measure small electrical currents.

When she met Pierre, Curie realised that her life had changed forever. After the ill-fated affair with Kazimierz, she had devoted her life to study but Pierre swept her off her feet. She saw in Pierre a kindred spirit and before long the two became inseparable. Pierre was different from any man she had known; intelligent and quiet, he loved science as much as she did. Just like Curie's parents, Pierre's family had placed a huge emphasis on education but he had not followed a conventional path.

Pierre had been taught privately at home. At eighteen, he received his *licence ès sciences* from the Sorbonne and was offered a job as an assistant in the Sorbonne's teaching laboratories. Soon he began to publish original work. Pierre had never obtained a PhD although he had done enough original work to have been awarded several. He was considered an outsider. In 1893 Pierre left the Sorbonne to teach at a new industrially oriented school, the École Municipale de Physique et Chimie Industrielles (EPCI).

Pierre was equally bowled over by Curie. He saw in her a startling intelligence. Their friendship quickly deepened and soon Pierre told Curie that he wanted to marry her. Curie loved Pierre but after her heartbreak with Kazimierz did not want to be hurt a second time. In summer 1894 she decided to leave Paris for Warsaw, feeling that she wasn't prepared to marry. She felt a sense of duty both to her family and to Poland.

Pierre was not prepared to give up on this remarkable woman. He wrote, begging her to return to Paris. He even offered to leave France and move to Poland. It is possibly this offer, more than anything else, that made Curie understand Pierre's feelings. She returned to Paris to commit herself to Pierre and a life in France. However, she did not take up his offer to share an apartment and instead found an apartment on rue de Châteaudun, next to Bronia's office.

Pierre had promised that on her return to Paris they would spend as much time as possible together but his mother fell ill and he had to stay with her in Sceaux in the south of Paris. One week he wrote to Curie, 'My father [a physician] has rounds to make and I will stay in Sceaux until tomorrow afternoon so that Maman won't be alone . . . I sense that you must be having less and less esteem for me while at the same time my affection for you grows each day.'

Pierre decided finally to write up his doctoral thesis and submit it to the Sorbonne, probably at Curie's insistence. In March 1895, about a year after they met, Pierre was examined on his thesis at the *faculté des sciences*, passing without issue. Soon after obtaining his PhD, a professorship was created for him at the EPCI, and he and his brother Jacques were awarded the *prix Planté* for their work on piezoelectricity, the production of electricity by putting pressure on crystals.

Later that spring, Curie made the final commitment. The pair married on 26 July 1895 at Sceaux town hall, with the reception in the garden of the nearby Curie family home. Curie's father and sister Helena came from Warsaw and her sister Bronia came with her husband. The Curies used wedding money from a cousin to buy two new bicycles and set off for Brittany on honeymoon, cycling from one fishing village to the next. 'We loved the melancholy coasts of Brittany and the reaches of heather and gorse,' Curie wrote.

Returning to Paris after a long honeymoon, the couple took an apartment on the rue de la Glacière, near to where Curie had lived as a student. By this time their combined income from salaries, prizes, commissions and fellowships was about 6,000 francs, three times a schoolteacher's salary. This allowed for a comfortable lifestyle but they were not extravagant and did not employ servants.

Curie returned to her work on magnetism and continued to study science and mathematics in her own time. She took two courses, one with Marcel Brillouin, a theoretical physicist with a wide range of interests. Pierre, meanwhile, was given his first class to teach at the EPCI, a course on electricity. According to Curie's later account, the course was 'the most complete and modern in Paris'.

From the start, the couple worked together as often as they could. French mathematician Henri Poincaré once said that their relationship was not just an exchange of ideas but also 'an exchange of energy, a sure remedy for the temporary discouragements faced by every researcher'.

In early 1897 Curie discovered that she was pregnant and suffered frequent dizzy spells and sickness. She was often unable to work and to add to her low spirits she learnt that Pierre's mother was terminally ill with breast cancer. Curie feared that her child's birth would coincide with Pierre's mother's death and was worried about the effect this might have on her husband.

By summer, seven months into her pregnancy, Curie left for Brittany to rest while Pierre stayed in Paris to finish teaching his courses and look after his ailing mother. When his brother Jacques took over caring duties, Pierre joined Curie in Port-Blanc on the northern coast. Despite Curie being very pregnant they both went cycling again. Shortly after they returned to Paris, Curie went into labour and on 12 September 1897 gave birth to a daughter, Irène. In Curie's meticulous household expenses book, she noted that they bought a bottle of wine to celebrate.

The expenses book also shows a jump in monthly expenditure on employees from 27 francs in September to 135 francs in December. Curie had employed a nurse and a wet nurse for Irène. Then, just as Curie feared, barely two weeks after Irène's birth, Pierre's mother died. Pierre's father moved in with his son, daughter-in-law and new granddaughter.

By now, Curie had obtained the qualification required to be a teacher and in late 1897 was assembling charts and photographs for her article on the magnetism of tempered steel for the bulletin of the Society for the Encouragement of National Industry. She and Pierre decided that she should begin original research and prepare for a doctorate.

The previous two years had been exciting in physics. In 1895, Wilhelm Röntgen discovered X-rays, leaving physicists scrambling to understand this strange new phenomenon. And in 1896, in what should have been an X-ray experiment, Henri Becquerel uncovered what were first known as Becquerel rays. Curie decided that she would study these new rays for her doctorate, as hardly anything about them was understood.

All that was known was that Becquerel rays were emitted by uranium and that they penetrated paper, making some materials glow in the dark. But, the study of these 'uranium rays' had lost momentum. Only a handful of papers were presented on the subject to the French Academy in 1896, compared with nearly one hundred on X-rays. Uranium rays were considered part of the phenomenon giving rise to X-rays; no one realised that they were caused by different processes.

With Pierre's help, Curie set up a laboratory in an old storeroom on the ground floor of the EPCI building. It was cold and dirty but Curie was happy to devote herself to a research topic of her choosing. She began her laboratory notebook on 16 December 1897. She and Pierre built an ionisation chamber to measure the energy given off by the uranium. These measurements were extremely tricky – Becquerel himself had failed – however, Curie believed that with care and diligence she would succeed and she did.

She began her work with the intention of submitting a PhD dissertation in which she measured quantities more accurately than before, with no expectations of new discoveries. After testing uranium and measuring the tiny electrical charges in the mysterious Becquerel rays she scavenged other elements to test. On one day alone in February 1898 she tested thirteen elements, including gold and copper, finding that none gave off uranium rays.

Had Curie stuck to testing pure elements, she would have missed the discovery that made her famous, but on 17 February she tried a sample of the black, heavy mineral compound pitchblende. This had been mined for over a century from the mineral rich Joachimsthal region on the German-Czech border. In 1789 Martin Heinrich Klaproth had extracted a grey metallic element from pitchblende, which he named 'uranium' after the newly discovered planet Uranus.

As uranium was a relatively small component of pitchblende, Curie expected the rays from the pitchblende to be weaker than those from pure uranium. To her amazement, she found the reverse. Initially she thought she had made a mistake but the result was confirmed after checking. But why was the radiation from pitchblende stronger? Curie tested other substances and a week later made another unexpected discovery. The mineral aeschynite, which contains thorium but no uranium, was also more active than uranium. Now she had two puzzles to solve.

Curie suspected that the rays that Becquerel had discovered were not just a phenomenon of uranium but something more general. Just as her research was moving in a new and unexpected direction, Pierre heard that he had been turned down for a professorship at the Sorbonne. The couple seemed to not dwell on this disappointment and in the laboratory notebook Pierre's handwriting appears with increasing frequency alongside Curie's as they worked more closely together.

The obvious assumption was that there was another, more energetic, element in the pitchblende, also giving off rays. But what could this element be? Pitchblende contained a mix of minerals, too many to duplicate in the laboratory.

About this time, the Curies discovered that another uranium-containing mineral, chalcite, also gave off more energetic rays than pure uranium. Chalcite was simpler to synthesise than pitchblende and the Curies reasoned that, if they created chalcite from its known ingredients, the mystery element would be missing and the rays would be less energetic. Curie mixed up artificial chalcite by combining copper phosphate and uranium and found that the new mineral showed no greater activity than uranium. The conclusion was clear – chalcite and pitchblende contained an additional unknown element. Curie wrote up her findings as *Rays Emitted by Uranium and Thorium Compounds*. It was read to the Academy on 12 April 1898. Neither Curie nor Pierre was a member of the Academy and only members could speak, but luckily Curie's professor, and by now good friend, Gabriel Lippmann, was willing to present the paper. The Academy members were intrigued by Curie's findings but did not pick up on two points in the paper which, in retrospect, were most important.

Curie had conjectured there was a new element in the pitchblende and chalcite that was responsible for the increased energy of the emitted rays. This introduced a novel technique for detecting new substances – that the radioactive properties of a material could indicate their presence. Secondly, in the paper, Curie states, 'All uranium compounds are active . . . the more so, in general, the more uranium they contain.' Implicit in this statement is the suggestion that the rays were an atomic property – an idea that turned out to be prophetic. But the Academicians were not convinced of the existence of a new element.

The only way to prove this would be to isolate the new element, which meant dealing with Becquerel. Curie and Pierre were in a difficult position. Becquerel had helped them get money to set up their laboratory and in many ways was a friend. But Curie felt slighted by the way that Becquerel always dealt with Pierre. Not only did she feel that he treated her as inferior but Becquerel took her ideas and conducted similar experiments in competition to her work.

Pierre tried to reassure his wife that Becquerel was not a rival but Curie was far more driven than her husband. She didn't want to share her discoveries with anybody else and certainly not a member of the Academy, a group of men who, she felt, looked down on female scientists. She was determined to beat Becquerel to the discovery of the new element.

A few days after her paper was read, Curie and Pierre were back in their laboratory, pulverising 100 grams (0.22 pounds) of pitchblende to try to isolate the mystery new element. They treated the pitchblende with various chemicals, measuring the activity of the products of chemical reactions. The most active of the breakdown products was then worked on further. Two weeks after starting, the Curies felt that they had isolated enough active product to determine its atomic weight using spectroscopy. This would show conclusively if they had found a new element.

Frustratingly, the substance showed no unknown spectral lines – the bright lines in the light spectrum that act as a fingerprint for identifying elements. This could have meant that Curie was mistaken in her belief in the existence of a new element, but she felt that they had not isolated enough. The Curies asked one of the EPCI's laboratory chiefs, Gustave Bémont, if he could help them in their chemical separation and purification of pitchblende. His input was immediately successful. Beginning by heating a fresh sample of pitchblende in a glass tube, he distilled a product that was strongly active. By early May they had a substance more active than pitchblende.

By now, the laboratory notebook suggests that Curie and Pierre were dividing their efforts. Very soon they had a product that was seventeen times more active than the uranium that they used as a benchmark. By 25 June, Curie had a product that was 300 times more active than uranium. Pierre, working in parallel, isolated a product 330 times more active.

The Curies were now beginning to think that pitchblende contained not one, but two new elements. One seemed associated with the bismuth in the ore and the other with barium. Feeling they had isolated enough of the element associated with bismuth, they tried spectroscopy again, calling upon spectroscopy expert Eugène Demarçay, but no new signature was found. Despite this lack of evidence, the Curies were convinced that the bismuth harboured an unknown element. On 13 July Pierre wrote a significant entry in the notebook. It is the first indication that they had given their hypothetical element a name – 'Po' – an abbreviation of 'polonium', the name the Curies chose in honour of Marie's home country.

Five days later Henri Becquerel presented a paper on their behalf to the Academy. 'We have not,' they admitted, 'found a way to separate the active substance from bismuth,' but they had 'obtained a substance which is 400 times as active as uranium.' They went on:

We thus believe that the substance we have extracted from pitchblende contains a metal never before known, akin to bismuth in its analytic properties. If the existence of this metal is confirmed, we propose to call it polonium after the name of the country of origin of one of us.

This paper was the first to introduce the term 'radio-active' in its title *On a New Radio-Active Substance Contained in Pitchblende*. Soon the term radioactivity was everywhere, leaving Becquerel rays, uranic rays and uranium rays abandoned. This notebook ends around the time of the paper on polonium and it seems that they did not get any further for three months. This may have been because they were waiting for a new shipment of pitchblende, but also probably reflected the customary departure of academics from Paris for several months during the *grandes vacances* in the summer.

Once they returned to work, progress was rapid. By the end of November 1898 they had isolated a highly active product, carried off by barium. With Bémont's help, they increased the radioactivity of this product to 900 times that of uranium. This time spectroscopy expert Demarçay found what they had hoped for – distinct spectral lines that could not be attributed to a known element. In late December, Pierre wrote the name for this second new element in the middle of the page of their notebook – radium.

One step remained to prove beyond doubt that they had discovered another new element – to isolate it and measure its atomic weight. For several weeks they compared the masses of samples of barium containing radium to the normal element, but they could not detect a difference in mass. They conjectured that this was because radium was present in tiny quantities. At the end of December they sent off their next paper to the Academy.

Entitled *On a New, Strongly Radio-Active Substance Contained in Pitchblende*, it was written by the Curies and Gustave Bémont. It included the report from Demarçay on his spectral analysis which not only picked up a new spectral signature but noted that those spectral lines 'intensify at the same time that the radioactivity intensifies . . . a very serious reason for attributing them to the radioactive part of our substance.' Demarçay further added that the spectral lines 'do not appear to me to be attributable to any known element . . . [its presence] confirms the existence, in small quantity, of a new element in the barium chloride of M. and Mme Curie.'

This announcement marked a turning point in the Curies' work pattern. Rather than continuing to work together on the same project, they set up in parallel. In early 1899, Curie took on the task of isolating radium while Pierre, working in the same laboratory, attempted to understand the nature of radioactivity better. Pierre took on the physics, while Curie handled the chemistry. Curie had a stubborn desire to isolate a sample of radium. She also knew that to win over the sceptics they had to isolate their new elements.

Curie had to resort to near-industrial methods, requiring a bigger laboratory. The Curies asked the Sorbonne but all the university could offer was an abandoned building formerly used as a dissection laboratory. The huge space had no heating, so in the winter was horribly cold. Pierre and Curie huddled around a small stove to keep warm, then hurried into the cold parts of the laboratory to conduct their work. There were no extractor hoods to carry away the poisonous gases given off by Curie's chemical treatment, so work had to be done in the courtyard. If the weather did not allow this, they worked inside, opening the windows.

By spring 1899 Curie had the materials she needed. As she later recounted:

> I had to work with as much as 20 kilograms [44 pounds] of material at a time . . . so the hangar was filled with great vessels full of precipitates and of liquids. It was exhausting work to move the containers about, to transfer the liquids and to stir for hours at a time, with an iron bar, the boiling material in the cast-iron basin.

Fairly early in the process, it became clear that it would be easier to separate radium from barium than polonium from bismuth. Despite the arduous work and long hours, Curie thrived on the challenge. As she worked on isolating the radium, the Curies had a unexpected source of delight – concentrated radium compounds were spontaneously luminous. Sometimes, after supper, the couple would wander back to their laboratory to admire the eerie glow of their samples. They sent small amounts of radium to fellow scientists around the world.

Unaware of the dangers, the Curies brought home radium salts in a glass jar, keeping it next to their bed. As the months went on, Curie, Pierre and Becquerel noticed the damage the radioactive materials were causing. Becquerel carried a glass tube of radium salts in his jacket; a few weeks later he found that his skin was burned next to the radium.

Pierre's work was also progressing. He reported the effect of magnetic fields on radium emissions and the Curies published a stream of papers. At the 1900 International Congress of Physics in Paris, they presented their longest paper yet, *The New Radioactive Substances*, in which they summarised their findings, plus work from England and Germany.

By this time it was known that some of the rays could be deflected by a magnet while others could not. Some rays penetrated thick barriers that others could not. And radioactive elements could 'induce' radioactivity in other substances – turning the Curies' lab radioactive. But no one knew how any of this worked. As the paper said, 'The spontaneity of the radiation is an enigma, a subject of profound astonishment.'

Isolating radium from pitchblende was exhausting and time-consuming. During this long undertaking Curie presented two progress reports in the journal *Comptes Rendus* in November 1899 and in August 1900. Then, finally, in 1902, Curie announced that she had successfully isolated one decigram (one tenth of a gram) of radium chloride. Her paper announced that the measured atomic weight of radium was 225, close to the current agreed value of 226, and concluded that 'according to its atomic weight, [radium] should be placed in the Mendeleev [periodic] table after barium in the column of the alkaline earth metals'.

Curie's isolation of radium was not only a huge achievement for sheer doggedness, it was crucial in developing our understanding of radioactivity. Physicist Jean Perrin noted in 1924, 'It is not an exaggeration to say today that [the isolation of radium] is the cornerstone on which the entire edifice of radioactivity rests.'

Curie went on to write up the work for her thesis and submitted it to the Sorbonne. In May 1903 she was awarded a doctorate. The day of celebration also resulted in a chance encounter with Ernest Rutherford (No 9 on our list, see chapter six), who was in Paris with his wife. Rutherford had called to see Paul Langevin, a fellow research student from the Cavendish Laboratories in the mid-1890s. Langevin invited the Rutherfords and the Curies to dine with him. After dinner, Rutherford recalled, they went outside into the garden, and Pierre 'brought out a tube coated in part with zinc sulphide and containing a large quantity of radium in solution. The luminosity was brilliant in the darkness and it was a splendid finale to an unforgettable day.'

In August, just two months after receiving her PhD, Curie suffered a miscarriage. Her grief was compounded when she learnt that Bronia's second child had died of meningitis. Curie fell ill with anaemia; it would be months before she could return to work. But by November the couple's fortunes began to change dramatically. On 5 November they learnt that the Royal Society of London had awarded them the Humphry Davy medal, given annually for the most important discovery in chemistry. Curie was too unwell to travel so Pierre went to London alone. On his return he found a letter from the Swedish Academy informing the Curies that they, together with Henri Becquerel, had won the 1903 Nobel Prize for physics.

The prize was in its infancy. The first physics prize was awarded in 1901 to Röntgen for the discovery of X-rays. The 1903 prize was to be awarded to Becquerel 'for his discovery of spontaneous radioactivity' and to the Curies 'for their joint researches on the radiation phenomena discovered by Professor Henri Becquerel'. There had been some discussion in the Swedish Academy over whether radioactivity fell within the purview of physics or chemistry. In the end it was considered physics and, to avoid pre-empting a possible future prize in chemistry, the discovery of radium was not mentioned. Curie became the first woman to receive a Nobel Prize and – until her daughter Irène was awarded a Nobel in 1935 – was the only female science laureate.

Pierre thanked the Swedish Academy but went on to say that neither he nor Curie would attend as they had teaching obligations, important research to conduct and Curie was not feeling well. Becquerel, on the other hand, went to Sweden, making little mention in his acceptance speech of the Curies' work. Despite this, the newspapers made a great deal of Curie as the first woman to win the prize. The Curies hated the publicity but several good things came out of it. The Sorbonne offered Pierre a professorship and Curie was given a better laboratory. Pierre was also made a member of the French Academy of Sciences.

Meanwhile, more radium was being isolated and the world was falling in love with this strange substance. It was assumed that a substance which gave off such a lovely glow must be good for you and radium quickly became a cure-all. Some bought radium-enhanced water while radium salts were sewn into performers' costumes to glow in the dark. In Paris, a Montmartre revue was entitled *Medusa's Radium*, and in San Francisco, USA, a production featured 'fancy unison movements by eighty pretty but invisible girls, tripping noiselessly about in an absolutely darkened theatre and yet glowingly illuminated in spots by reason of the chemical mixture upon their costumes'. Radium was painted onto the faces of watches and clocks and one company even brought out a radium lipstick. As yet, no one realised the harm the substance could do but already Curie and Pierre were feeling its ill effects.

Pierre's hands became so damaged from handling radium that he had difficulty dressing himself. His bones ached and he walked like a man twenty or thirty years older. Curie too was often weak. It seems strange that neither made any connection between their deteriorating health and the radiation with which they worked. In a paper Pierre wrote during this period he noted that laboratory animals breathing the emanations of radioactive substances in a confined space died within a matter of hours. The paper concluded, 'We have established the reality of a toxic action from radium emanations introduced into the respiratory system.'

Despite Curie's frequent bouts of weakness and ill health, in December 1904 she gave birth to her second daughter, Ève. The Nobel Prize had come with 70,700 Swedish kronor, the equivalent of around £290,000 ($450,000) in 2015 and the Curies started taking more holidays with their children. They bought better clothes and Curie sent a substantial part of the fund to her family in Poland. It seemed as if she had finally found personal and professional happiness.

This was shattered on a rainy day in April 1906. Pierre had gone to a meeting and was walking to the Sorbonne. Hampered by the effects of radium, Pierre paused while crossing a busy Parisian street and a large, horse-drawn carriage hit him. Despite the driver's best attempts, the carriage wheel ran over Pierre's skull, crushing it instantly. Curie returned home from work to learn the tragic news; she was, not surprisingly, devastated.

It was work that slowly helped Curie put aside her grief. The Sorbonne offered her Pierre's professorship, making her the first female professor at the university. Her first lecture, on 5 November 1906, was scheduled to begin at 1.30 p.m., but several hundred people had gathered in front of the Sorbonne's iron gate well before midday to see this historic event. When the doors opened, the crowd rushed in, filling every seat and standing in the aisles. If winning the Nobel Prize made her a celebrity, her determination to carry on with her work led to France taking Curie to their hearts.

She decided to move out of Paris and into the countryside, away from the apartment which carried too many memories. There she arranged private tutors for her daughters. It was already clear that Irène was showing signs of following in her parents' footsteps, exhibiting an early aptitude for science and mathematics. Ève, on the other hand, loved music.

Soon it became known to Curie's friends that she had fallen in love again. The man in question was Paul Langevin, Pierre's former student. However, Langevin was married and his wife discovered love letters from Curie. The wife threatened to kill this other woman. Curie begged Langevin to get a divorce but he was not prepared to break up his family. He promised his wife that he would not see Curie again, except in a professional capacity. That same year, Curie was nominated to become the first female member of the French Academy of Sciences. Yet when the votes were cast in January 1911, she was refused. Her friends were outraged but Curie shrugged it off.

In November 1911, Curie attended the Solvay conference in Brussels, a meeting for the main players in physics including Einstein, Rutherford, Becquerel, Röntgen and Langevin. Langevin's wife suspected his promise to end the affair was a lie. In a rage, she took Curie's love letters to the newspapers, precipitating a scandal. The day after Marie returned to Paris, the headlines of *Le Journal* carried the front-page headline A STORY OF LOVE: MADAME CURIE AND PROFESSOR LANGEVIN.

The love affair dominated French newspapers for days, almost overshadowing the news that Curie had made history by being awarded a second Nobel Prize, this time in chemistry. The citation was 'for the discovery of the elements radium and polonium by the isolation of radium and the study of the nature and compounds of this remarkable element'. So great was the infamy of the Langevin story that the Nobel committee wrote to Curie asking her to refuse the prize. Curie wrote back to say that her private life had nothing to do with the quality of her research. She would accept the prize, and in person. In December 1911 she received her Nobel from the King of Sweden, bringing her sister Bronia and Irène to the ceremony.

On 29 December Curie was rushed into hospital and for the next two years suffered from a severe kidney ailment. For most of January 1912 she was under the care of the Sisters of the Family of St Mary on rue Blomet, but after returning home her health did not improve and in March she went back to hospital for an operation.

By this time she weighed only 47 kilograms (103 pounds), 9 kilograms (20 pounds) less than she had three years earlier. She wrote to the dean of the Sorbonne asking for time off and was too ill to return to teaching for another six months. She also suffered increasing bouts of what we now know to have been radiation sickness, but at the time the cause of these bouts was a mystery. Though her affair with Langevin was over, the scandal had taken its toll and the public who had once adored her did not forgive her readily. She was the scarlet woman who enticed Langevin into adultery. People threw stones at her windows and newspapers continued to snipe at her. Curie disappeared from public view for several years.

To stay abreast of developments in her field, Curie travelled under a false name, leaving her daughters with a governess. Gradually, the press lost interest in the scandalous Madame Curie and moved on. Curie eventually found that, once again, she could move about Paris without hostility.

In the summer of 1914 Irène passed her *baccalauréat* and planned to enter the Sorbonne herself. She and her mother were becoming partners as Irène continued to excel in science – the subject she wanted to study at university. Both assumed that they would one day work together in a laboratory but world events were to intervene. The French government announced the construction of a dedicated centre for Curie's research but the Radium Institute – later renamed the Curie Institute – was opened just as World War I broke out.

With Paris under threat, the government announced in August 1914 that 'the radium in the possession of Mme Curie, professor of the faculty of sciences of Paris, constitutes a national asset of great value'. On 3 September Curie took all the radium that France had accumulated in a lead-lined case to Bordeaux and hid it in a vault in the university. As she travelled back she learned that the German army had retreated and the Battle of Marne had begun. French forces, reinforced by soldiers sent to the front line in taxicabs from Paris and the British Expeditionary Force, overwhelmed the Germans and Marne was won. Paris was safe – for the time being.

Curie realised that the war could provide new purpose and opportunities – in times of crisis, even the moral transgressions of Sorbonne professors seemed trivial. It could allow her to put the pain of the affair behind her. She also saw a way to help her native Poland, which had become the battleground between Russia and Germany.

Some sixteen days after the war had begun, the Russian Tsar announced that he intended to give Poland its autonomy. In a letter to *Le Temps*, Curie described this as 'the first step towards the solution of the very important question of Polish unification and reconciliation with Russia'.

As was characteristic of Curie, her devotion to the war effort was huge and complete. The notebooks in which she noted every expense show many entries for charitable donations. There are entries for aid sent to Poland, for national French aid, for 'soldiers', 'yarn for soldiers' and shelters for poor people. In addition, according to her daughter Ève's account, Curie invested the money from her second Nobel Prize in French war bonds, which became essentially worthless. She even tried to contribute her medals, but officials at the Bank of France refused to melt them down.

Curie finally hit upon a way to use her expertise to help out after a conversation with the eminent radiologist Dr Henri Béclère. He told her that X-ray equipment was scarce and 'when it existed was rarely in good condition or in good hands'. Curie became determined to make X-rays available to wounded soldiers at or near the front. Although she was not a radiologist, she knew how to make X-rays; she wrote to Irène, 'My first idea was to set up radiology units in hospitals, employing the equipment that was sitting unused in laboratories or else in the offices of doctors who had been mobilised.'

This hospital work served as a good training ground; she learnt the rudiments of X-ray examination from Béclère and passed this knowledge to volunteers she recruited. However, these visits to hospitals made her realise that what was really needed was a mobile post that could carry an X-ray machine and all the associated equipment. Curie found benefactors to donate a car – the French Red Cross and the *Union des Femmes de France*. Finally, she needed to find the necessary equipment.

'This little car . . . carrying only the equipment that was strictly necessary, has without doubt left many memories in the Paris region,' she said after the war. 'Manned in the beginning by voluntary personnel, former students at the *École normale* or professors . . . it alone provided service to the troops retreating to Paris during the greater part of the war, in particular at the time of the flood of wounded who arrived in September of 1914 following the Battle of the Marne.'

In October a second car was donated and Curie decided to try to persuade the army to give her radiology cars official backing. After weeks of her request passing from desk to desk, on 1 November, the minister of war granted permission for her radiology cars to go to the front lines. Curie, Irène, the mechanic Louis Ragot and a chauffeur set off in radiology car number two for the Second Army's evacuation hospital at Creil, 32 kilometres (20 miles) from the front line at Compiègne. Curie made a remarkable contribution to the war effort, equipping eighteen radiology cars that examined ten thousand wounded soldiers. By 1916 she had obtained a driving licence and so, when necessary, even drove herself.

Curie also trained and educated others, realising that equipment without the training to go with it was useless. She was asked by the army to conduct a course for X-ray technicians but, following months of difficulties with the military approach, she decided to train nurses instead. She opened a school for female radiologists in October 1916 and between its opening and the end of the war the school trained about 150 women. They completed a six-week course and were sent to radiology posts around the country.

Curie's recollections of this period were published in her book *Radiologie et la Guerre*. The greatest praise in her book was for Irène, who worked closely with Curie throughout the war and became, at the age of only eighteen, a teacher on the course for female X-ray technicians. This partnership, forged during the war, continued in the laboratory throughout the rest of Marie's life. By September 1916, Irène was working on her own as a radiologist in Hoogstade, a small part of Belgium that remained unoccupied by Germans. Incredibly, Irène also managed to obtain her certificates from the Sorbonne, passing all subjects with distinction; mathematics in 1915, physics in 1916 and chemistry in 1917.

In the Treaty of Versailles of 1919, Poland became a sovereign nation for the first time in 123 years and Curie wrote, 'A great joy came to me as a consequence of the victory obtained by the sacrifice of so many human lives.' After the war, Curie wanted to help heal the wounds that had arisen in the scientific community. She was asked to join the commission on intellectual cooperation of the League of Nations and served for over twelve years.

She also encountered the ambitious American journalist, Marie Meloney, who wrote to Curie asking for an interview. Expecting to find this great woman of French science installed 'in one of the white palaces of the Champs-Élysées', Meloney found herself face to face with 'a simple woman, working in an inadequate laboratory and living in a simple apartment on the meagre pay of a French professor'. Meloney decided that Curie needed her help and Curie sensed an opportunity to acquire some of the radium which the USA had accumulated.

According to Meloney, Marie said, 'America has about fifty grams of radium. Four of these are in Baltimore, six in Denver, seven in New York.' She went on to name the location of every grain. She then added that her own laboratory 'has hardly more than a gram'. Meloney quickly realised that it would help a great deal if she could get enough money together to donate a gram of radium to Curie's laboratory. The journalist set about attracting the funds, in the process portraying Curie as 'impoverished', which was far from the truth.

By June 1921 Meloney's fundraising mission was largely accomplished when she managed to raise over one hundred thousand dollars to buy a gram of radium. Meloney had arranged for Curie to visit the USA in May with opportunities to give lectures, accept honorary degrees and to be presented with a gram of radium by President Harding at the White House. Curie did not want to travel so early in the year and she pushed for an October visit. Meloney wrote to the rector of the Academy of Paris, encouraging him to put pressure on Curie.

In the end, Curie compromised, leaving Cherbourg on 4 May 1921 with her daughters on RMS *Olympic*. Meloney had by this time committed Curie to a ten-week stay during which she would attend many luncheons and dinners and award ceremonies with a little time off to visit Niagara Falls and the Grand Canyon. Large crowds gathered at the pier to welcome Curie in New York and so began a tour that Curie found both exhilarating and exhausting. She was treated with great enthusiasm like a celebrity, much as Einstein had been a few years earlier – at least in most places.

Despite many colleges and universities awarding Curie an honorary degree, the physics department of Harvard voted not to do so. When Meloney asked retired Harvard president Charles Eliot why, he replied that physicists felt that the credit for the discovery of radium did not belong entirely to Curie and that she had done nothing of great importance since her husband died. Harvard did welcome her warmly, so Curie herself probably had no idea of these behind the scenes machinations.

On 20 May 1921, Marie Curie attended a reception in the Blue Room of the White House. President Warren G. Harding presented Curie with the key to a green leather case containing an hourglass with the 'symbol and volume of one gram of radium' (the actual radium was safely stored in a laboratory). Curie's response was brief; she was tired after the hectic tour and had had to cancel several engagements due to fatigue. At times Irène and Ève found themselves receiving honorary degrees or medals on their mother's behalf.

The press was rife with speculation about Curie's fatigue. It was said that the small talk was too much for her, that she was unaccustomed to socialising and she was not used to leaving her laboratory. But the main source of her illness was undoubtedly her long exposure to radiation. Marie herself said privately during her tour, 'My work with radium . . . especially during the war, has so damaged my health as to make it impossible for me to see many of the laboratories and colleges in which I have a genuine interest.' The Curies made a sweep of the west before boarding the *Olympic* to return to France and Curie's beloved *Institut du Radium*.

The *Institut* came out of the desire of the Pasteur Institute and the Sorbonne to build a laboratory for work on radioactivity. After some infighting, a deal had been made to build two separate institutes. One – funded and run by the Sorbonne with Curie as director – was dedicated to the study of the physics and chemistry of radioactive elements; the other focused on the medical applications of radioactivity. The second was funded and run by the Pasteur Institute and directed by Dr Claudius Regaud, a medical researcher from Lyon. The two buildings had been built side by side.

From the time that it opened in 1914, Curie's laboratory employed a remarkably high number of women. In 1931, twelve out of thirty-seven researchers were women. In 1939, Marguerite Perey, working in Curie's laboratory, discovered the element francium and became the first woman to be elected to the Academy of Sciences, fifty-one years after Curie had been rejected.

The risks of working with radiation became more apparent both in and out of the Institute in the years following World War I. In 1925, a young woman named Margaret Carlough who painted luminous watch dials in a factory in New Jersey, USA, sued her employer, the US Radium Corporation. She claimed that her work, which involved using her lips to point her brush, had caused irreparable damage to her health. As the lawsuit progressed it came to light that nine dial painters from the same factory had died and it was concluded that these deaths were due to radiation. By 1928, fifteen dial painters had died from exposure to radium.

In Curie's laboratory, the effects of radiation were also beginning to take a toll. In June 1925 engineers Marcel Demalander and Maurice Demenitroux died within four days of one another from exposure to radioactive materials for medical use. A radiologist had to undergo a

Figure 5: the Radium Institute, circa. 1934. On the left is the Curie Laboratory, directed by Curie. On the right is the Pasteur Laboratory, directed at the time by Dr Claudius Regaud.

'series of amputations, of fingers, of his hand, of his arm' while another worker lost his eyesight and several others died after terrible suffering. In November 1925, Irène received a letter from the Japanese scientist Nobus Yamada, who worked closely with her preparing polonium sources at the Institute, to say that he had fainted two weeks after returning home and since then had been confined to bed. Two year later Yamada was dead.

Although she tried to deny it, it was clear to both Curie and those close to her that her own health was deteriorating. She went to extraordinary lengths to hide it. 'These are my troubles,' she wrote to her sister Bronia. 'Don't speak of them to anybody, above all things, as I don't want the thing to be bruited about.' By the early 1920s her eyes had become weak and she had a near continuous humming in her ears. According to her daughter Ève, Curie went to incredible lengths to keep her poor eyesight secret. She placed coloured signs on her instruments and wrote her lecture notes in huge letters. As Ève wrote, 'If a pupil was obliged to submit to Mme Curie an experimental photograph showing fine lines, Marie by hypercritical questioning, prodigiously adroit, first obtained from him the information necessary to reconstruct the aspect of the photograph mentally. Then and then alone she would take the glass plate, consider it and appear to observe the lines.'

In all, Curie had three cataract operations and told Ève, 'Nobody needs to know that I have ruined eyes.' But even if radiation was ruining her health, Curie did not want to retire. As she said to Bronia in a letter in 1927, 'Sometimes my courage fails me and I think I ought to stop working, live in the country and devote myself to gardening. But I am held by a thousand bonds . . . Nor do I know whether, even by writing scientific books, I could live without the laboratory.'

Curie returned one more time to the USA, in 1929, to fulfil a promise, as a group of American women had raised enough money to buy another gram of radium for a new institute in Curie's native

Poland. This institute eventually opened in 1932. Curie was presented with a cheque for the radium bound for Poland by President Hoover but apart from seeing a few friends she was too ill to repeat the visits of her previous tour.

Curie's health was deteriorating fast and in January 1934 she joined Irène and her husband Frédéric Joliot on a trip to the mountains of the Savoie. Over Easter she made her last visit to her house in Cavalaire with Bronia. Curie went down with bronchitis and had to cut the holiday short. After five weeks convalescing, she returned to Paris where Ève was waiting for her. She was increasingly suffering from fevers and chills.

From May, Ève saw a rapid decline in her mother. Doctors in Paris saw old tubercular lesions on an X-ray and suggested that Curie be taken to a sanatorium in the Savoie Alps. On the train, Curie fainted but when she was carried to a bed in the sanatorium doctors found no evidence of tuberculosis. A Swiss doctor who examined her blood found 'pernicious anaemia in its extreme form'. At dawn on 4 July 1934, in the peaceful sanatorium in the clear Savoie mountain air, Marie Curie died from the steady accumulation of radiation in her body.

While Curie had been the first person to isolate highly radioactive new elements, the next person in our list, Ernest Rutherford, would unravel the mechanism that ultimately killed her.

....

CHAPTER SIX

Ernest Rutherford

....

The solid No 9 on the list is New Zealander Ernest Rutherford, whose team split the atom. Rutherford was nothing like the archetypal introverted scientist. A loud presence with a robust sense of humour, he infamously categorised science by saying, 'There is physics and everything else is stamp collecting.' His aim was to emphasise how competing disciplines like biology, chemistry and astronomy were primarily engaged in cataloguing rather than developing fundamental theories, making it ironic that his Nobel Prize, awarded in 1911, was for chemistry.

At a time when most scientists worked in isolation or in small locally nurtured groups, Rutherford was renowned for finding the best people in the world to work with and for bringing on bright new talent. If he did not know something, he would find someone who did and work with them. His career took him from a farm in New Zealand to head up the best physics laboratory in the world – the Cavendish at Cambridge.

Rutherford was born on 30 August 1871 in Spring Grove, New Zealand. Spring Grove (now Brightwater) is a small town 12 miles (20 kilometres) south-west of Nelson, towards the north coast of the South Island of New Zealand. Rutherford's father, James, was from Perth in Scotland. He emigrated to New Zealand as a five-year-old, on a ship that took six and a half months to reach this new but familiar-feeling land.

James Rutherford was a farmer, but by no means rich. This outpost of the British empire was a land where people worked hard to scrape by. As a consequence, a strong work ethic was instilled in young Rutherford. Before and after school he was expected to help on the farm; there were no long lazy summer holidays for the Rutherford household who were put to use in numerous tasks to ensure the farm returned a profit, albeit meagre.

Rutherford's mother, Martha Thompson, had also emigrated to New Zealand as a child, with her family from Hornchurch in Essex, England. Before marrying James, Martha had been a teacher with a significantly higher level of education than her future husband. This was exemplified by her owning that iconic symbol of frontier gentility, a Broadwood piano.

Rutherford was the second son and fourth child of James and Martha's eventual brood of twelve. His mother recalled that he read his first science book aged ten – *Primer of Physics* by Balfour Stewart, which his mother saved long after he had read it from cover to cover. He attended the local state primary school, Havelock, and was the best student this small school had ever seen.

At fifteen, Rutherford took the scholarship examination to attend Nelson Collegiate School and, such was his local standing, that a small crowd gathered to see how their local hero got on. He won the scholarship, recording the highest ever score of 580 out of 600. At Nelson, Rutherford won many prizes and scholarships in subjects including English, history, French and Latin. But it was in mathematics and the sciences that he excelled.

At eighteen, Rutherford won a scholarship to Christchurch College, Canterbury, part of the University of New Zealand. He received his BA in 1892 and was awarded a scholarship to do a year's postgraduate study. In 1893 Rutherford graduated with an MA in physical science, mathematics and mathematical physics. He was then persuaded to undertake another year of research at Christchurch and obtained a BSc at the end of 1894 with his thesis *Magnetisation of Iron by High-Frequency Discharges.*

Rutherford then applied for an 1851 Exhibition scholarship to do further postgraduate study at the Cavendish Laboratory in Cambridge, England. These scholarships are a remarkable legacy of the Great Exhibition at the Crystal Palace in Hyde Park in London. The first ever world trade fair had been a huge attraction, pulling in more than five million visitors in its five-month run. When the exhibition closed in October 1851, a royal commission was established to spend the profits in improving science and industry.

The commissioners first purchased eighty-seven acres of land in South Kensington where they helped with the construction of the remarkable collection of buildings that include the Victoria & Albert Museum, the Science Museum, the Natural History Museum, the Albert Hall and Imperial College, a site still largely owned by the commission. There was enough money left to set up an educational trust which started the Science Research scholarships in 1891 and still awards twenty-five postgraduate fellowships and scholarships a year.

Rutherford was not the only New Zealander to apply, contesting a single place with chemist J. C. Maclaurin. Rutherford lost out to the older man and it looked like he would have to find another career path. But at the last minute Maclaurin got engaged and, as a married man could not take up the award, the message came through to the Rutherford home where the young scientist was working in the garden. Allegedly, Rutherford threw his spade into the air, shouting, 'That's the last potato I'll ever dig.' He became the Cavendish's first research student.

The Cavendish Laboratory, set up under James Clerk Maxwell – No 5 in our list – had come under the directorship of Joseph John (J.J.) Thomson in 1884 and by the 1890s was the world's leading experimental physics laboratory. Rutherford followed in Maxwell's footsteps by investigating the way that a magnetised needle lost some of its magnetism when placed in a magnetic field produced by an alternating current. He spent the next year improving the sensitivity of his instrument so that it was able to detect electromagnetic waves from a distance of 800 metres (half a mile), a record at the time. What he had effectively produced was a detector of electromagnetic waves.

However, the English scientific tradition separated research and technology and it was frowned upon for a Cavendish-based scientist to seek a commercial application of his discoveries. Guglielmo Marconi was doing similar work and while Rutherford was ahead of him it was the Italian who would commercially develop radio (then known as wireless) technology. Rutherford was encouraged to stick to 'pure' research.

Thomson decided to establish a group to investigate the new phenomena of X-rays and radioactivity, discovered by Röntgen and Becquerel respectively. Rutherford was one of his first choices, for his considerable talent and raw ambition to succeed. This is illustrated in a letter Rutherford wrote to his fiancée Mary, back in New Zealand, in October 1896 in which he stated:

> I have some very big ideas which I hope to try and these, if successful, would be the making of me. Don't be surprised if you see a cable some morning that yours truly has discovered a half a dozen new elements.

Thomson's idea was to use X-rays to investigate the production of invisible ions – atoms which have either gained or lost an electrical charge. The word 'ion' comes from the Greek word for wanderer, as ions were seen to wander in an electric or magnetic field due to their electrical charge. The details of their nature were a mystery and Thomson hoped to uncover their properties. Rutherford loved this work because the paths of ions could be followed; it allowed him to investigate their motion and energy in a way which would otherwise not have been possible. It was the first glimpse into an atomic-scale world, and Rutherford referred to the ions as 'Jolly little buggers . . . One can almost see them.'

Thomson decided to concentrate on investigating the charge-to-mass ratio of the most common ions. If he could find some relationship between electrical charge and the mass of the ions it might give a better picture of the structure of an atom. This led Thomson in 1897 to discover the first sub-atomic particle, the electron, for which he received the Nobel Prize in 1906. Meanwhile, Rutherford started to look into ultra-violet radiation and then to investigate the radiation produced by uranium – the process Marie Curie termed radioactivity.

Rutherford discovered that radioactive emission was more complex than had been initially thought. He found that some radiation from uranium was easily blocked by thin foil whereas another type often penetrated the same foil easily. It appeared there were two different types of radiation and Rutherford called them alpha and beta rays. He showed that both types were deflected by electrical and magnetic fields, indicating that they were charged particles. The alpha particles were less able to penetrate than the beta particles by about a factor of one hundred. Becquerel discovered that Rutherford's beta particles had the same charge to mass ratio as Thomson's electrons, making it highly likely that they were the same.

By the time he came up with the terms alpha and beta radiation in 1899, Rutherford had left Cambridge to take up a professorship at McGill University in Montreal, Canada. He had been recommended for the position by Thomson and, although Rutherford knew that McGill was a research backwater, he was keen to move on from Cambridge as he was increasingly resentful of his treatment.

Rutherford had found that the young elite of Cambridge University were openly hostile to a colonial from a distant land. Rutherford felt he had been badly treated throughout his time in

Cambridge. He was never offered the position of fellow at his college, Trinity, which would have provided him with an apartment and a lifetime stipend of £350. He didn't share his hurt with any of his colleagues, but in a letter to his fiancée Mary he wrote:

> As far as I can see my chances for a fellowship are very slight. All the dons practically and naturally dislike very much the idea of one of us getting a fellowship and no matter how good a man is, he will be chucked out . . . I think it would be much better for me to leave Cambridge on account of the prejudice of the place . . .

An added incentive was that McGill had recently been endowed with one of the best equipped laboratories in the world, thanks to the generosity of Montreal millionaire William McDonald. Strangely for someone who had made his fortune from tobacco, McDonald referred to smoking as 'a disgusting habit', whereas Rutherford was a keen smoker. Whenever McDonald visited, Rutherford would rush around the lab, opening windows and making sure all traces of smoking were eliminated.

At McGill, Rutherford expanded his study of radioactivity to thorium, one of the few elements then known to be radioactive. Realising that he needed a chemist, Rutherford employed Frederick Soddy, then a demonstrator in the chemistry department, to help him. Between 1900 and 1903, Rutherford and Soddy developed a theory to explain radioactive decay in terms of atoms emitting alpha and beta particles from within and in doing so changing to different elements. This was known as the 'transformation' or 'disintegration' theory.

Rutherford and Soddy claimed that the energy for radioactivity came from within the atom and that when an alpha or beta particle was emitted, the source transformed into a different element. The new theory was not too dissimilar from the age-old alchemist's idea of elements changing from one to another, a notion that had been long discredited. But such was the weight of Rutherford and Soddy's experimental evidence that their theory quickly gained widespread acceptance.

Another significant concept was that the radioactivity of thorium decreased steadily over time. Rutherford and Soddy noted that it was reduced by half in sixty seconds and by a further half in the next sixty seconds. They formulated an exponential decay law, linking a set fraction of the radioactive material to a set period of time of decay. This led Rutherford to the idea of half-life, a term he coined in 1907. Half-life is the time taken for half of the atoms in a sample to undergo radioactive decay. For different radioactive elements this can range from fractions of a second to billions of years. We now know the half-life is unique for each radioactive element or isotope.

Penetrating radiation discovered by French scientist Paul Villard in 1900 was identified by Rutherford three years later as a third type of radioactive decay, which he termed gamma rays. Rutherford demonstrated that the rays were not deflected by a magnetic field, unlike alpha and beta rays, showing they were not charged particles. This was confirmed in 1910 by William Henry Bragg, who showed that gamma rays were a form of electromagnetic radiation like X-rays. In 1914, Rutherford and co-worker Edward Andrade measured the wavelength of gamma rays and found them similar to but shorter than X-rays.

By 1904, Rutherford had begun to explore the possibility that radioactive decay was not a single process, but rather followed through a decay series or decay chain of products, starting with radium, decaying to radon which then went through a range of intermediate products. He gave the products names from radium-A (in fact polonium 218) to radium-F (polonium 210). Through a series of beta and alpha emissions, the end product was stable lead.

Three years later, Rutherford collaborated with American chemist Bertram Boltwood at Yale and the pair discovered that radium was a decay product of uranium. The end product of this decay chain (now called the 'radium' or 'uranium' chain) was a stable isotope of lead and the other two decay chains – the thorium chain and the actinium chain – also ended in stable isotopes of lead. Rutherford and Boltwood showed that by measuring the slowly increasing amount of lead in a radioactive mineral, they could establish that rocks were billions of years old. This was an important step in proving the age of the Earth. Geologists had suggested similar ages for geological processes and the figure also supported the time span required for Darwin's idea of evolution by natural selection.

One famous anecdote from Rutherford's days at McGill involved his bumping into geologist Frank Dawson Adams one day while walking across the university campus. According to the story, Rutherford asked Adams, 'How old is the Earth supposed to be?' to which Adams replied, 'One hundred million years.' Rutherford then reached into his pocket and pulled out a lump of pitchblende and said, 'I know for a fact that this piece of pitchblende is seven hundred million years old,' and walked off, chuckling.

Rutherford then set about studying radioactive particles in more detail. He felt that the alpha particle, which was much more massive than the beta particle, was key in his transformation theory. He was able to show that it carried a positive charge but was not able to determine whether it was a hydrogen ion (a single proton) or the larger helium ion (we now know it is the latter).

As Rutherford's work became better known, the job offers started pouring in. He received lucrative offers from Yale, Columbia and Stanford. He was even offered the directorship of the Smithsonian Institution in Washington. But Rutherford longed to return to the more vibrant scientific scene in Europe. In 1906 the chance came. Arthur Schuster, Langworthy Professor of Physics at Victoria University (now the University of Manchester), the second most important laboratory in England after the Cavendish, sent him a letter stating that he planned to retire and would be relieved if

Rutherford would succeed him in his position at Manchester. The news of his leaving McGill even made the *New York Times*, which commented in January 1907:

McGill University is about to suffer a severe loss through the resignation of Prof. Ernest Rutherford, McDonald Professor of Physics since 1898. He has accepted a chair in Victoria University, Manchester, England.

Prof. Rutherford, who is only 35 years old, stands in the front rank among the physicists of the world through his research work on radium, as well as his previous studies and discoveries in connection with wireless telegraphy.

In 1908, soon after his arrival at Manchester, Rutherford's work on radioactivity was honoured with the Nobel Prize for chemistry. The citation read, 'For his investigations into the disintegration of the elements and the chemistry of radioactive substances.'

On arriving in Manchester, Rutherford started working with German physicist Hans Geiger on an ion detector; the Geiger counter became the universal instrument for measuring radioactivity. With student Thomas Royds, Rutherford was able to isolate some alpha particles and show that they were helium ions. Rutherford then undertook the research for which he is most famous.

Continuing his long-standing interest in alpha particles, he studied the way they were scattered when fired at thin gold foil. Geiger placed his counter in different positions to see how the alpha particle count varied as a function of angle to the incoming beam.

In 1909 an undergraduate, Ernest Marsden, joined the experiment for his research project. Rutherford suggested that Marsden see if any alpha particles were deflected back towards the same side of the gold foil from where they were fired rather than passing through. Much to everyone's surprise, such deflections of more than 90 degrees were detected, leading Rutherford in 1911 to suggest that atoms have dense, positively charged nuclei at their centres.

Rutherford's theory of atomic structure took a while to gain widespread acceptance but in 1913 the Danish physicist Niels Bohr (see chapter eight), after visiting Rutherford's laboratory, suggested that radioactivity could be explained by energy changes in the atomic nucleus and that, in addition, the spectra of gases – the colours in the light given off when they are heated – could be explained by only allowing electrons to orbit the nucleus in particular orbits. These suggestions led to a wider acceptance of Rutherford's model.

Breaking for war work, in 1917 Rutherford returned to working on the effects of shooting alpha particles into various gases. He noticed that alpha particles fired into nitrogen gas produced hydrogen ions. He suggested that all atomic nuclei consisted of particles which were the same as the hydrogen ion. He had in fact produced the first nuclear reaction, changing normal nitrogen (nitrogen-14) into oxygen-17 (an isotope of normal oxygen) and a free hydrogen ion. It was in 1920 that Rutherford himself named the hydrogen ion particle the proton. He also postulated the existence of a neutral particle in the nucleus, a neutron, something that was experimentally confirmed at the Cavendish in 1932 by James Chadwick, brought by Rutherford from Manchester.

With the retirement of J. J. Thomson in 1919, Rutherford was offered and accepted the directorship of the Cavendish Laboratory. He spent the rest of his career there, working on nuclear reactions.

The Cavendish team found that it was easier to bombard nuclei with neutrons than with charged particles, because they were not electrically repelled by the nucleus. In 1934, Rutherford, Mark Oliphant of Australia and Paul Harteck of Germany bombarded deuterium, a naturally occurring form of hydrogen with a neutron in its nucleus, to produce tritium, a new radioactive form of hydrogen with one proton and two neutrons.

Rutherford, knighted in 1914 and made Baron Rutherford of

Nelson in 1931, died in Cambridge in 1937 following a short illness and was buried in Westminster Abbey. His discovery of the nucleus changed our understanding of the atom, but the next person on our list changed our understanding of the very nature of space and time and became the most recognisable scientist ever.

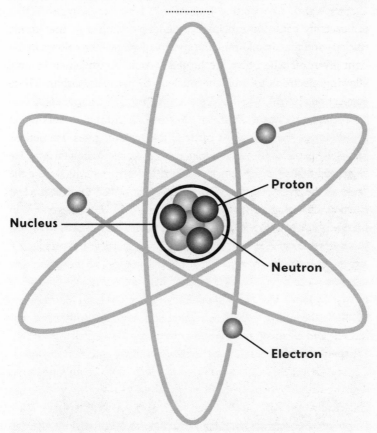

Figure 6: the Rutherford model of the atom, in which most of the mass of the atom was in a central nucleus (now known to consist of both protons and neutrons), and the much less massive electrons were in orbit about the nucleus. For obvious reasons, it is often referred to as the 'Solar-system' model.

CHAPTER SEVEN

Albert Einstein

....

At a surprising No 4 in our list is Albert Einstein, the man who revolutionised our concept of space and time and replaced Newton's law of gravity with a completely different interpretation of the force.

Einstein's name is synonymous with genius and yet as a student he barely passed his degree in physics. Then, in a remarkable series of papers in 1905 written while working as a patent clerk, he not only overthrew the cherished ideas of Newtonian mechanics but explained Brownian motion and the photoelectric effect. In the same year, Einstein's PhD thesis on the diffusion of gases gave the best evidence to date of the existence of atoms and in his fifth paper that year he came up with the most famous equation in physics – $E=mc^2$. Einstein went on to become the world's first celebrity scientist and was instrumental in persuading President Roosevelt to develop the atomic bomb. Along with Newton, Einstein towers over the world of physics.

Albert Einstein was born in Ulm, Germany on 14 March 1879. His father, Hermann, ran an electrochemical enterprise but soon after Einstein was born, with the business failing, the family moved to Munich. Hermann's business had been funded by money from Einstein's mother Pauline's family and Hermann's brother Jakob persuaded him to move to Munich to go into business together.

Einstein was so late starting to talk that his parents were concerned that he was backward. But when he finally did speak he formed carefully constructed, full sentences. In November 1881, when Einstein had just turned two-and-a-half, his sister Maja was born. His mother had promised him 'something new to play with', so when Maja was born he was disappointed and asked where the wheels were.

Generally, Einstein was a quiet child, though as a young boy he could have a violent temper and when agitated would throw anything to hand at the nearest person, usually Maja. Once, he smashed a garden hoe over her head and when Pauline decided it was time for him to learn the violin he had such a fit of rage that it scared off the first teacher. Thankfully, a more resilient teacher helped Einstein develop a lifelong love of playing the instrument.

By the standards of the day, Hermann and Pauline's approach to parenting was unusual as they encouraged an extraordinary amount of independence. When Einstein was only four he was encouraged to make his own way around the streets of their suburb. The following year, Einstein started school. Although his parents were Jewish they were not particularly religious and sent the boy to a nearby Catholic school. Einstein hated it; it depended on rote learning with raps over the knuckles for any mistakes – a world apart from the freewheeling atmosphere at home.

In the same year, Einstein fell ill and was in bed for several weeks. To relieve the boredom, his father gave him a compass to play with and Einstein became fascinated by it. He was enthralled that, no matter how he turned it, the needle always pointed north. What was the mysterious force that was doing this? Einstein found it far more interesting than anything he learned in school, and this self-discovery made a lasting impression.

When he was ten, Einstein was enrolled at the Luitpold gymnasium. Although bored by most subjects, he had developed a love of mathematics. This was partly due to his uncle Jakob, but mainly thanks to a medical student, Max Talmud, who joined the Einstein family for supper every Thursday. Talmud was studying in Munich away from home and at the time it was a tradition amongst Jewish families to help young students, providing home comforts during their times away. Jakob introduced Einstein to algebra while Talmud discussed the latest things that he was learning at university. Later in life, Einstein would say that these twin influences were the most important factors in his becoming a scientist.

Sadly, the teaching at the gymnasium was as dull as the primary school. Einstein soon lost interest and his marks slipped steadily. Teachers commented that he was a lazy boy who would never amount to anything. The impending crisis came to a head in 1894 when Einstein was fifteen. With the business in Munich also failing, Jakob persuaded Hermann that their financial future would be better in Italy. The family sold up and emigrated but because Einstein was at a crucial stage in his schooling he stayed in Munich where he lodged at a boarding house.

This independence at fifteen helped mature the young Einstein and was also probably a factor in his own attitude to parenting when he later had children. He still found his classes dull and within six months of his parents leaving he persuaded the family doctor that he was on the brink of a nervous breakdown. He took the doctor's letter to his mathematics teacher, the only teacher he got on with, who gave him a letter saying that there was nothing more that he could teach him.

With these two letters, Einstein went to see the gymnasium's headmaster and announced that he was leaving the school. The headmaster replied that, on the contrary, he was expelling Einstein for being such a poor student and so a fifteen-year-old Einstein headed south to join his family in Pavia, Italy.

While wanting to be with his family was undoubtedly behind this move, there was possibly another reason too. Einstein had developed deep pacifist feelings, a conviction which he would keep throughout his life. In Germany, upon turning seventeen, all males were required to do a year's military service. The thought horrified the young Einstein; by moving to Italy he was also escaping this dreadful prospect.

Einstein arrived unannounced in Pavia and later Maja described her brother as being in high spirits; there was no sign of the 'nervous breakdown'. Soon after arriving, Einstein renounced his German citizenship and persuaded Hermann to allow him to spend several months touring the art centres of Italy before he settled down to real work.

When he returned from this tour he and Hermann were at loggerheads over the future. Einstein's prospects did not look very rosy. He had vague ideas of becoming a teacher but Hermann wanted him to serve an apprenticeship and go into the family business. Whichever path he was to take, he would need a college diploma and so Einstein applied for a place at the prestigious Federal Institute of Technology in Zurich (known as the ETH after its German name *Eidgenössische Technische Hochschule*). In the autumn of 1895, six months before his seventeenth birthday, Einstein set off to take the ETH entrance exam.

The normal age of admission was eighteen but Einstein was confident that he would gain entry early. He was shocked when he failed; he did well in mathematics and science but in the other subjects his scores were very poor. The authorities, looking upon him favourably because of his age, told him that if he enrolled at a Swiss school and obtained a high school diploma he could gain entry to the ETH the following year without retaking the exam.

Although disappointed, Einstein realised that he had no choice. Fortunately, his father Hermann found him a school in a small town called Aarau which had a relaxed atmosphere and approach to learning that suited Einstein perfectly. His enjoyment of the year was also helped by meeting Marie, the daughter of the school's headmaster, Jost Winteler. Einstein boarded with Professor Winteler and his family, living in close proximity to Marie where the two quickly fell in love. The relationship did not survive the following year when Einstein went to Zurich but his sister Maja later married Marie's brother Paul.

Einstein applied himself to his studies at Aarau. When he took the diploma examinations he sailed through, obtaining perfect marks in history, geometry and physics. He didn't do so well in the other subjects, but passed them all and on 29 October 1896 he arrived in Zurich to embark on the next chapter of his life.

The freedom given to ETH students appealed greatly to Einstein. He attended the lectures he liked and didn't bother with the others – which turned out to be most of them. His free time was spent in cafés and bars, engaged in animated discussions with his friends on everything from politics to literature and physics. One of his lecturers, the mathematician Hermann Minkowski, described the young Einstein as 'a lazy dog who never bothered about mathematics at all'.

In Einstein's year at the ETH there were only five students on the science course. One was a woman, Mileva Maric, and Einstein quickly established a close friendship with her. By his last year at the ETH between 1899 and 1900 their relationship had developed into a full-blown love affair. As their final exams approached, Einstein realised that he had to get down to some serious studying of the material that was actually on the syllabus as opposed to the things he found interesting. He borrowed the notes of fellow classmate and close friend Marcel Grossman and furiously set about catching up.

Of the four students who graduated, Einstein achieved the lowest scores. The only student to do worse was Mileva, who failed. Having barely scraped a degree, Einstein now had the difficult prospect of finding a job. Although he had dreamed of teaching at a university, the professors at the ETH were not going to offer Einstein a position. Indeed, none of his professors was likely to give him a good reference. When he hadn't been absent he had been constantly challenging his lecturers and disagreeing with their approach to physics problems.

Einstein's wish was to carry out independent research which, he hoped, would lead to his being awarded a PhD. With his theoretical bent, this was in principle possible. As long as he could access a good library, he could work without being enrolled at any university. But Einstein needed to support himself financially during the several years it would take to produce a thesis worthy of a doctorate.

Meanwhile, Mileva decided that she would retake her final year. A few months before she was due to resit her exams she discovered that she was pregnant. Not surprisingly, she hardly welcomed this news and failed her final exams a second time. By the end of 1901, Einstein found himself a temporary post as a teacher. Things looked very bleak for the couple but thanks to Einstein's friend Marcel Grossman, a ray of hope appeared. Grossman persuaded his father to have a word with a friend, the director of the Swiss patent office in Bern, about a job and through this connection Einstein was made an offer of employment; he started work in June 1902.

Mileva went into hiding with her parents so that her pregnancy would not be noticed, as the scandal of an illegitimate child would ruin any prospects for Einstein to have a career in academia. At the end of January she gave birth to a daughter, Lieserl, who was given up for adoption and never heard of again. Even today, no one knows what became of her.

Einstein would stay at the patent office until July 1909. Although it seems an unlikely job for a budding theoretical physicist, the work suited Einstein perfectly. His sharp mind was able to quickly see any flaws in the patent applications he reviewed and he did not find the work demanding. This allowed him time to think about physics.

A couple of months after starting this position, in August 1902, his father died. With Hermann's death, the objections to Einstein marrying Mileva disappeared and the couple got married on 6 January 1903. Einstein had gone from near disaster to a good job and a respectable family life.

Mileva increasingly found herself in the role of housewife and Einstein often enjoyed the company of a group of friends including Maurice Solovine, Conrad Habicht and Michelangelo Besso. This group, calling themselves the Olympia Academy, met frequently to discuss new ideas in physics. They played a key role as a sounding board when Einstein started to develop the ideas that would see their fruition in his 'miraculous year' of 1905.

Before this year, Einstein had published a few papers and while none was remarkable they were solid pieces of work. One recurrent theme was Einstein's attempt to establish the reality of atoms and molecules, a hotly debated issue. This question provided the motivation for his PhD thesis that he submitted in 1905. In this thesis Einstein calculated the way that molecules behaved in solution. Specifically, he studied sugar molecules dissolved in water, calculating the way that the sugar would exert pressure within water and comparing his calculations to experimental data from the literature.

As part of his work, Einstein calculated the size of a sugar molecule, making it to be a little more than a millionth of a centimetre across. This calculation was elegant and groundbreaking. Einstein finished the work in April and submitted his thesis to the ETH in July but was told by the authorities that it was too short. Later in life Einstein would take great delight in telling people that he added one sentence to his thesis before resubmitting it; this time it was promptly accepted.

The delay between completion and submitting his thesis in 1905 was due to preoccupation with other work. On 11 May a paper by Einstein arrived in the offices of German journal *Annalen der Physik*. The paper was on Brownian motion – this is the strange jiggling of tiny particles (such as pollen grains) when they are in a liquid or floating in air. Some suggested that the motion was due to the particles being hit by molecules, either of water or air, but missed the idea that each zig or zag of a pollen grain was due to it being hit by an individual molecule. If molecules were causing this motion, they had to be comparable in size to the pollen grains.

As Einstein had determined the size of molecules for his thesis work, he applied this to Brownian motion and showed that the effect would occur if there was constant, but not always equal, bombardment of each pollen grain by water or air molecules. Crucially, Einstein put precise calculations into his theory, and described the zigzag motion rigorously in statistical terms. It was this mathematical rigour that persuaded most physicists that his theory was correct.

This was the second paper that he submitted to the journal that year; his March 1905 paper on the photoelectric effect would later win him the Nobel Prize. Einstein was explaining a phenomenon that had first been noticed experimentally towards the end of the nineteenth century. When light was shone onto the surface of some metals an electric current was produced.

The most detailed investigations of the effect had been carried out in 1902 by Hungarian scientist Philipp Lenard. Lenard found that by changing the wavelength of light on the surface, the effect could be made to go away. The effect was more often there in blue light, but there was no current with red light. In addition, a brighter light led to a larger current (more electrons being liberated in the metal) but each electron was measured to have the same energy, which only depended on the wavelength of the light, not on its intensity.

Physicists had pondered the nature of light for centuries. Newton's corpuscular theory (see chapter two) had been displaced by light travelling as a wave, an idea championed by Christiaan Huygens. An experiment conducted by Thomas Young in 1800 seemed to resolve the issue. Young found that when light from a single source passed through two narrow slits an interference pattern was produced – a pattern that could only be explained if light travelled as a wave. Then, in 1865, when Maxwell published his equations describing electromagnetism (see chapter four), he argued that light was an oscillating wave in an electric field and magnetic field at right angles to each other.

This was the view accepted in the early 1900s. However, a few chinks had emerged. The first was when German physicist Max Planck, in 1900, tried to explain blackbody radiation. This is the radiation given off by a substance that absorbs all incoming radiation. The surface of the Sun and all stars approximate to blackbodies and their temperatures are sufficiently high that the radiation given off is visible to our eyes. We also give off blackbody radiation because of the temperature of our bodies, but our bodies have much lower temperatures and the blackbody radiation is in the infrared part of the spectrum.

It was known from experiments conducted in the 1890s that the radiation from a blackbody at different wavelengths followed a particular curve. Going from longer to shorter wavelengths (or lower to higher frequencies) the radiation would rise to a peak and then drop off again. The position of these peaks allowed a precise calculation of the temperature of the blackbody to be made. However, the theory to explain this curve did not work – the theory predicted that the curve would keep on rising and rising, emitting an infinite amount of energy. This failure of the theory, called the ultraviolet catastrophe, was one of the biggest problems in physics at the end of the nineteenth century.

Max Planck looked for a mathematical equation to fit the curve. He succeeded, but had to allow only specific frequencies to emit radiation. He had to assume that light could only be emitted in chunks which he called 'quanta'. The energy of each was given by $E=hf$ (where f is the frequency of the light and h is a constant which we now call Planck's constant). Most physicists found Planck's solution ad-hoc, and even he regarded it as nothing more than a method of calculation – he did not think quanta existed. But in 1905, Einstein extended Planck's work to argue that the photoelectric effect could be explained if light was composed of actual quanta.

Einstein argued that each light quantum ('photon', as we call them today) carries an energy given by Planck's equation and that a photon could either give all or none of its energy to an electron at the surface of the metal. If the photon's energy was sufficient it could liberate the electron but as its energy was dependent on the frequency of the light, blue light could work where red light would not. In addition, Einstein argued that the intensity of the light merely increased the number of photons hitting the surface each second, not the energy of each photon. This explained what Lenard had found – the energy of the electrons did not depend on the intensity of the light.

At the time of publication, Einstein's paper made little impact. This was partly due to his position as an outsider but mainly because physicists were unwilling to accept that light was anything but a wave. American Nobel-winning physicist Robert Millikan was strongly opposed to this idea; to prove Einstein wrong he conducted a series of careful experiments. As he later remarked:

> I spent ten years of my life testing that 1905 equation of Einstein's and contrary to all my expectations, I was compelled in 1915 to assert its unambiguous verification in spite of its unreasonableness.

The papers on the photoelectric effect and Brownian motion, plus his PhD thesis, were just the start of the work in 1905. On 30 June another paper arrived at the *Annalen der Physik* titled *On the Electrodynamics of Moving Bodies*, concerning the special theory of relativity.

As we saw in chapter one, Galileo introduced the idea of relativity when he argued that there was no mechanical experiment that could distinguish between being at rest or moving with uniform motion. Maxwell had suggested that light was a manifestation of electromagnetic waves because of the coincidence he found between the speed of light and the speed of his waves. But for light to be a wave it required a changing electric field to produce a changing magnetic field, and the changing magnetic field would then produce a changing magnetic field and so on. Einstein thought about what would happen if you could travel alongside a light wave. This would make the electric field static, producing no magnetic field. The light would disappear.

In addition, the speed of light in a vacuum is fixed in Maxwell's equations; according to Newton, however, it should vary depending on your own motion. If light travels at 300,000,000 metres a second (186,000 miles a second) and an observer travelled towards it at 100,000,000 metres a second (62,000 miles a second), Newton's laws argued that the observer should measure that light at 400,000,000 metres a second (248,000 miles per second) – but this is not what Maxwell's equations predicted. Either Maxwell or Newton was wrong and most assumed it was Maxwell.

Einstein believed it was Newton who was not correct. He took Galilean relativity, which only applies to mechanical experiments, and extended it to light – hence the title of the paper. Einstein argued that there was no experiment, mechanical or using light, which could distinguish between being at rest or moving with a constant velocity.

Beginning with two postulates – the extended idea of Galilean relativity and the speed of light being constant – Einstein showed that our concept of space and time had to be altered. Space and time became relative when these requirements were plugged into the equations of motion. Two observers travelling relative to each other would measure time and space differently. However, these effects only become evident if the relative speed of the two observers was about half the speed of light, which is why we never notice the difference in everyday life. The fastest a human has travelled with respect to the Earth was on Apollo 10 at 39,896 kilometres per hour (24,790 miles per hour) – around 0.000037 times the speed of light.

Special relativity becomes apparent at truly high speeds, with results that can include time dilation and length contraction – which are opposite sides of the same coin. Time dilation means a moving object's time runs slowly when seen from the place that it is moving with respect to. This results in a shrinking of the length of the moving object in the direction of travel – length contraction – and vice versa. The effect is symmetrical, so if a person in a high-speed rocket could see you, they would see time passing more slowly for you while you would see time passing more slowly for them. This is not just an apparent difference; it has been shown to be true thousands of times a day in particle accelerators. The particles produced in accelerators decay over short periods of time but because they are moving at high speed, they live for longer than they should.

Another prediction of special relativity is that the mass of an object increases as it gets faster and that nothing can travel faster than light speed. As we saw in chapter two, Newton's second law of motion, $F=ma$ (where F is force, m mass and a acceleration), tells us that if we apply a force to an object it will accelerate. Einstein showed that this no longer applies at high speeds because of increasing mass – as mass gets higher, the force needed to accelerate it increases. Again, this effect is seen every day in particle accelerators. As we speed up the protons or electrons in our giant atom-smashers, their mass increases in accordance with the predictions of Einstein's theory.

Not long after Einstein had sent off his paper he had an afterthought. He realised that his theory implied a connection between mass and energy, because mass changes as we increase an object's kinetic energy (its energy of motion). He wrote up this connection in a brief paper which appeared in the *Annalen der Physik* in November 1905 – a paper containing $E=mc^2$, the most famous equation in science. This equation shows that matter and energy are interchangeable. E is the energy, m the mass of the object and c the speed of light. As light speed is such a big number (300,000,000 metres per second; 186,000 miles a second), a small mass is the equivalent of a huge amount of energy – this is the principle by which the Sun gets its energy and an atomic bomb can release such devastating destruction.

These papers on special relativity went on to become iconic but at the time hardly anyone noticed them. Einstein's home life looked like that of any man in his mid-twenties. His first son, Hans, had been born in May 1904 and in September 1904 his salary at the patent office was increased from 3,500 to 3,900 Swiss francs; a comfortable salary enabling him and his family to take holidays. He took up sailing, a hobby that he maintained for the rest of his life.

Although Einstein's relativity papers did not gain much attention, his other output in 1905 led to his name becoming known among a small group of physicists, particularly in Germany. In April 1906 Einstein was promoted to technical expert (second class) with another pay rise. He could have settled down into a comfortable existence; his future at the patent office was secure and the salary was plenty to support his family. But his friends were urging him to seek a university appointment; he now had a PhD and his published papers showed an ability to think in an original way.

However, moving into academia was not trivial. He first had to find a position as a *privatdozent*, an unpaid lecturing position in a university. This allowed the lecturer to teach a subject of his or her choosing and students were charged a fee for attending. In 1907 Einstein applied to Berne University but was rejected, partly because the head of department said that he found special relativity 'incomprehensible'. Einstein looked into several other teaching jobs but nothing came of them until in 1908 he was finally successful with Berne, staying on at the patent office to keep a salary coming in. He proved an uninspiring lecturer, leaving his future in academia doubtful. Then, unexpectedly, he received an important boost from Hermann Minkowski, his former maths lecturer.

Soon after Einstein graduated, Minkowski had moved to Göttingen University as professor of mathematics. He was one of the first prominent academics to recognise the significance of Einstein's special theory of relativity and it was Minkowski who made the crucial step of linking Einstein's equations with a geometrical description in four dimensions. Minkowski coined the term 'spacetime' to highlight the idea that the two could not be separated. As Minkowski said in a lecture in Cologne on 2 September 1908:

The views of space and time which I wish to lay before . . . are radical. Henceforth, space by itself and time by itself are doomed to fade away into mere shadows and only a kind of union of the two will preserve an independent reality.

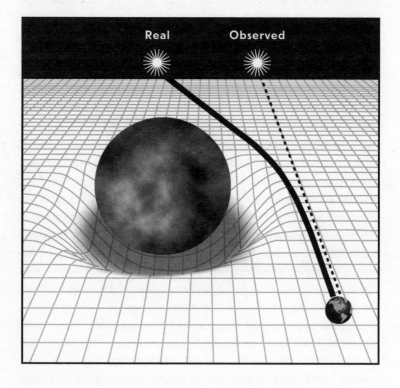

Figure 7: a diagram of the bending of starlight by the Sun, one of the predictions of Einstein's general theory of relativity. Because of this bending, a star would appear in a slightly different place in the sky ('Observed' rather than 'Real') when its light grazed the edge of the Sun. This prediction was spectacularly verified by Arthur Eddington in 1919 by making observations of stars during a Solar eclipse.

Initially, Einstein was irritated by Minkowski's apparent attempt to take over his idea and was unconvinced by Minkowski's attempt to put his equations into geometric form, but this proved to be the device that a wider audience needed to understand the theory. Soon after Minkowski's lecture, Einstein received the first offer of a paid appointment – as professor of physics at the University of Zurich. On 6 July 1909 he resigned from the patent office.

It was not long before Einstein was turning down offers as word of his work spread. But even before he left the patent office, Einstein had what he would later describe as 'the happiest thought of his life'. This was the idea that a person in freefall would not feel the effect of gravity and that, as they fell, other falling objects would float next to them. After this initial thought it did not take Einstein long to devise the principle of equivalence, suggesting that the effects of gravity and acceleration are indistinguishable. This idea would have far reaching implications for our understanding of gravity.

One consequence is that gravity should bend light. Imagine you are in a rocket which is in empty space and that rocket is either stationary or moving at a constant speed. As we have seen, Einstein's special theory of relativity said that there would be no way to distinguish between the two. Imagine a beam of light shot from one wall inside the rocket to the other. To both an inside and outside observer, the light would travel in a straight line.

Now, instead, suppose the rocket were accelerating. To an outside observer a light beam will not traverse the rocket in a straight line but instead will follow a curve. In fact, the curve will follow the very precise shape of a parabola, just as Galileo had shown (see chapter one). Through Einstein's principle of equivalence, what is true for acceleration is true also for gravity and hence Einstein predicted that light would be bent by gravity.

As Einstein considered his 'happy thought' in more detail he realised that it implied that gravity would also affect the passage of time. If one has two observers at each end of a rocket which is accelerating in space, one can show that the observer at the back of the rocket will measure time to be passing more slowly than the observer at the front. Through the principle of equivalence, this means that a person in a stronger gravitational field would measure time as passing more slowly than a person in a weak gravitational field – gravity slows down time.

Although this effect is subtle, we have to take account of the practical ways that it affects Earth's global positioning satellite (GPS) system. The satellites which orbit the Earth are in a weaker gravitational field than we are on the surface and the timings used to measure a position are so precise that they include the effect of time passing more quickly for the satellites than it does for us.

Einstein would go on to publish his new work on gravity as the general theory of relativity – 'general' because it tells us what will happen even if there is acceleration, while the 'special' theory only applies when velocities are constant. But Einstein had to delay publication for eight years while he learned an entirely new branch of mathematics in order to convert his thoughts on gravity into a full-blown theory. This new mathematics was the geometry of curved surfaces, developed by Bernhard Riemann and Carl Friedrich Gauss in the mid-1800s, but was completely alien to Einstein (and any other physicist at the time).

Returning to the geometrical interpretation of his special theory which Minkowski had highlighted, Einstein realised that the best way to describe how gravity and acceleration were equivalent was to argue that gravity shaped the very fabric of spacetime. As Princeton physicist John Archibald Wheeler once put it, 'Matter tells space how to curve and space tells matter how to move.' (Purists point out he should have said 'spacetime' rather than space.)

In the meantime, between his happiest thought in 1907 and publishing his general theory in 1915, Einstein's personal life was in flux. In July 1910 his second son Eduard was born and in 1911 he left the University of Zurich to take up a better position at the University of Prague. The following year, the ETH offered its wayward student a professorship taking him back to Zurich. Each year his fame was growing and almost every university in Europe was after him.

In 1914, Max Planck was dispatched by Berlin University to head-hunt Einstein at any cost. Planck offered Einstein a dream job – professor of his own research institute at Berlin with a huge increase in salary and no teaching requirements so that he could concentrate on research. Planck and Berlin wanted Einstein to be on their staff when his new theory of gravity was published.

The University of Berlin was the very heart of physics in the early part of the twentieth century. Despite Einstein's misgivings about returning to Germany, in the spring of 1914 he left Zurich for Berlin; but he was to pay a heavy price. His marriage to Mileva, already unstable, began to fall apart. She and the boys initially moved with him to Berlin but after a few months she took the children back to Switzerland, effectively signalling the end of their marriage.

At the university, Einstein insisted upon the freedom to work on whatever he wanted without outside interference. However, a few months after his arrival, the political landscape changed dramatically. With the outbreak of World War I, academics came under strong pressure to support Germany's war effort and most were happy to comply. Fritz Haber, Berlin's renowned chemist, worked on developing poison gases that would ultimately kill hundreds of thousands of allied troops and even Planck fully supported Germany's war effort. Along with ninety-two other distinguished scientists and public figures, Planck signed a 'manifesto to the civilised world' which put forward the argument that Germany was not the aggressor but rather defending what was true and proper. Einstein refused to sign, alienating him from his colleagues.

Einstein lived alone, sending money to support his children back in Zurich. Obsessed by his theory of gravity, he didn't eat properly and his health began to suffer. Einstein lost a lot of weight while most of his dark hair turned grey. Not far from his Berlin apartment his cousin Elsa lived with her two daughters. Recently divorced, Elsa took pity on Einstein and kept an eye on him, providing him with meals and trying to make sure his condition did not worsen. For Einstein, the obsessive working was worth it. He managed to generalise his special theory to include acceleration and hence to provide a completely new interpretation of gravity.

Einstein presented the general theory of relativity in a series of lectures to the Prussian Academy of Sciences in November 1915 and publication followed in 1916. Many people since have described it as a true landmark in physics in a much more fundamental sense than the special theory – had Einstein not developed the earlier special theory, others would have got there. Henri Poincaré, for example, was very close to time dilation in work he published in 1904. But the general theory of relativity took a much greater leap and we might still be waiting for such a theory if it were not for Einstein.

A scientific theory must describe all known phenomena that fall within its purview and make predictions of things which have not yet been observed. Einstein's theory passed the first test when it was applied to Mercury's orbit, which it had been known for decades was not as it should be. Like all the planets, Mercury orbits the Sun in an ellipse and the point at which it is closest to the Sun in its orbit is its 'perihelion'. Mercury's perihelion was moving (precessing) in space more like the series of ellipses one would produce in sketching the petals of a flower. Although some precession was expected from the gravitational influence of the other planets, Newton's law of gravity gave the wrong answer for what was observed.

The belief in Newton's theory was strong. After all, it had correctly predicted the existence of a new planet, Neptune, in the mid-1800s. Astronomers argued that Mercury must have an unseen companion named Vulcan which similarly affected its orbit. Einstein was unaware of this problem with Mercury's orbit but when it was pointed out to him he quickly calculated the precession predicted by his theory. It came out in triumphant agreement with observations. But there was more to come.

German scientific journals were banned in the UK because of the war so Einstein's theory of gravity had not reached British physicists and astrophysicists. However, in 1916, Willem de Sitter, a mathematician from neutral Holland, passed details to Arthur Eddington at Cambridge University, the most eminent young astronomer of his time. Eddington was an immediate convert to general relativity. (Famously, when later asked by a reporter whether it was true that only three people in the world understood the theory, he is said to have replied, 'Who is the third?') Eddington realised that one way to test Einstein's prediction that light would be bent by gravity would be to look around the Sun during a Solar eclipse. Light from stars which should be just behind the Sun during an eclipse should be bent, bringing them into sight. Eddington believed that taking photographs of those stars' positions and comparing those positions to a photograph taken six months earlier should enable him to test the theory. He knew that there was an eclipse due on 29 May 1919 that would sweep across part of Brazil, the Atlantic Ocean and on to the African continent.

Eddington raised enough money from the Royal Astronomical Society and the Royal Society to mount two expeditions, one to Sobral in Brazil and a second to the island of Príncipe off the west coast of Africa. On the day of the eclipse the skies were cloudy on Príncipe and it looked hopeless until they cleared just as the period of totality was about to end. Eddington was able to get a few photographic plates recording the stars' positions. Back in Cambridge he started the painstaking task of measuring the stellar positions on his plates precisely.

On 6 November 1919, at a joint meeting of the Royal Society and the Royal Astronomical Society, Eddington announced that Einstein was correct. (More recently it has been suggested that Eddington's evidence was borderline at best – but many experiments have since confirmed the result.) The announcement made it into the newspapers and overnight, at age forty, Einstein became a celebrity.

The years he had spent finishing the theory had taken their toll on Einstein, who had lost about twenty-five kilograms (sixty pounds) in weight. The indigestion that he frequently suffered turned out to be an ulcer and he had gone from a young-looking thirty-five to seeming much older than his forty years. For the latter half of 1917, he was largely bedridden and was looked after by his cousin Elsa. In the late summer of 1917 he moved into an apartment in the same block as Elsa, but his recovery was slow. It wasn't until almost a year later, in the spring of 1918, that he was able to leave the house.

Through the illness, Einstein and Elsa had grown increasingly close. By summer 1918 she wanted to put their relationship onto a more permanent footing. Einstein wrote to Mileva who agreed to a divorce, on the condition that she would get all the money from the Nobel Prize that Einstein was likely to receive. Einstein had been nominated for one every year since 1910 so it was hardly presumptuous to include this in the divorce settlement. At the time, the prize money was 30,000 Swedish kronor, then enough to provide a comfortable living for Mileva and their two sons. The divorce was finalised in February 1919 and in June Einstein married Elsa, who was three years older than him.

After 1915 Einstein made fewer significant contributions to physics. However, ever since his work on the photoelectric effect he had remained interested in the idea of light quanta and kept abreast of the developments in atomic theory. In 1916, Einstein made an important contribution to Bohr's concept of electrons having fixed orbits in atoms (see chapter eight) when he predicted an effect of the way that electrons could jump between these orbits or 'energy levels' as physicists call them.

Bohr's model required that an electron could jump to a higher level if it either collided with another electron or absorbed a photon, but that it would quickly try to return to a lower energy level. When it jumped back down, the electron would emit a photon of a frequency related to the energy difference between the two levels by the equation

$E=hf$ that Planck introduced in his 1900 paper on blackbody radiation. Einstein added the concept of 'stimulated emission'. Here, electrons which have been pushed up to a higher energy level are stimulated to jump back down by the passage of light of the correct frequency, effectively amplifying the light as it passes through. This is the basis of the laser – or Light Amplification by Stimulated Emission of Radiation.

Einstein was eventually given the Nobel Prize in 1922 for his work on the photoelectric effect – probably because his theories of relativity were still not properly understood by the members of the Swedish Academy. In the summer of 1924, Einstein had a burst of activity, making his last great contribution to physics.

It was prompted by his receipt of a letter from a young Indian scientist, Satyendra Bose. Bose showed Einstein a new way of deriving Planck's equation describing blackbody radiation without assuming that light behaves as a wave. He treated light as a gas of photons and used a new kind of statistics to describe them. Einstein refined and improved this idea, extended its application and the new system became known as Bose-Einstein statistics. They describe the behaviour of particles called bosons (the Higgs is the most famous boson thanks to CERN but photons are also bosons, as are the particles that carry the strong and weak nuclear forces). It was this work that led to the acceptance of photons in the physics community. (The term 'photon' was coined by American chemist Gilbert Lewis in 1926.)

By this time Einstein was one of the grand old men of physics and so was consulted on any new theory. He became a strong advocate of wave-particle duality, proposed by Louis de Broglie in 1923, referring to the idea in glowing terms in a paper he published in 1925. But, as de Broglie's ideas gave way to full-blown quantum mechanics, Einstein found himself increasingly outside the mainstream. As we shall see in chapter eight, Einstein was never able to come to terms with the statistical nature of quantum mechanics, famously protesting, 'God does not play dice.' By the end of the 1920s, Einstein was increasingly in disagreement with Bohr, the lead architect of quantum mechanics.

Einstein had another worry. As the Nazis became increasingly powerful, Einstein feared for his personal safety as a Jew. In 1931, American educator Abraham Flexner started work on a new centre for scientific research to which he hoped to attract the world's greatest scientists. The following year, Flexner visited Caltech (or the California Institute of Technology) to discuss the idea with Robert Millikan, the Nobel laureate who ran the establishment. Millikan hoped to entice Einstein to settle at Caltech and made the mistake of telling Flexner how much kudos a name like Einstein would bring.

A few weeks later, Flexner visited Einstein – then in a brief spell at Oxford – to try to persuade him to move to the USA and head up his new research centre. Einstein was intrigued but did not take the idea too seriously. Flexner, though, was persistent and a few months later he visited Einstein again; this time he got his man. Einstein would join the new Institute for Advanced Study that Flexner based in Princeton in the autumn of 1933, intending to divide his time between Berlin and Princeton.

On 10 December 1932, Einstein and Elsa left Germany for a stay at Caltech. Just under two months later, Hitler was appointed Chancellor in Germany and Einstein knew that it was no longer safe to live in Germany. During his time in Caltech during the winter of 1932–33 his German home was ransacked and his books were burned in public. When Einstein did return to Europe it was not to Germany; he visited Oxford and announced when visiting the Belgian seaside resort of Le Coq sur Mer that the threat of Nazism was so great that he could not go home. On 7 October 1933 he set sail for the USA to settle in Princeton, where he remained for the rest of his life.

In just over twenty years at the Institute, Einstein did not achieve any great breakthroughs, spending most of his time in a futile search for a theory to unite gravity and electromagnetism. However, his scientific influence never waned. When Eugene Wigner and Leó

Szilárd realised the need for the USA to develop an atomic bomb to defeat the axis powers, it was to Einstein they turned to write a letter to President Roosevelt to persuade him of the need for nuclear research.

Elsa died in 1936 and from then until his own death in 1955, Einstein's affairs were looked after by his secretary, Helen Dukas, who fiercely guarded his privacy. It was not until after her death in 1982 that many details of Einstein's private life became known, including the existence of his first child, Lieserl, who had been given up for adoption.

Soon after World War II, Einstein was asked to become the first president of the new state of Israel but turned it down. By now he had become the epitome of the absent-minded professor, the Einstein with crazy white hair who didn't wear socks because he forgot to put them on one day and afterwards decided he didn't need them. He remains the best-known scientist in history and one who completely overthrew our concept of space and time. But, despite his genius, Einstein never fully accepted quantum physics, the revolutionary theory derived from his work by the next person in our list – Danish physicist Niels Bohr.

....

CHAPTER EIGHT

Niels Bohr

Niels Bohr was the Danish theoretical physicist who gained a lifetime's free supply of lager from Carlsberg for his contributions to physics, probably the only physicist in history to receive such a perk. He was one of the few people to prove Einstein wrong in long-running mental jousts the pair conducted over randomness in quantum theory. Bohr comes in at No 2 on our list, a placing that many would think was remarkably high. However, his German theoretical physicist colleague Werner Heisenberg and others who helped develop quantum mechanics would have had no qualms about the position. As Heisenberg said:

> Bohr's influence on the physics and physicists of our [the twentieth] century was stronger than that of anyone else, even Einstein.

Bohr was born on 7 October 1885 in Copenhagen. His father, Christian, was a professor of physiology at Copenhagen University and was nearly awarded the Nobel Prize for his work on the chemistry of the respiratory system. Bohr's mother Ellen, of Jewish descent, came from a family of prominent bankers and politicians. Bohr was brought up in a household with a progressive-liberal-intellectual ethos, alongside his elder sister Jenny and his eighteen-month-younger brother, Harald. Both boys attended the prestigious Gammelholm gymnasium in Copenhagen. Harald was highly competitive; by the time they left school he had caught up and sometimes surpassed Bohr in most subjects. Witty and lively to Bohr's quiet introversion, Harald was an accomplished footballer and a gifted mathematician.

It would seem that despite the rivalry between the pair there was never any bad feeling. This is possibly because Harald pursued mathematics while Bohr concentrated on physics. This way, they could help and consult each other without competing. Jenny studied at Copenhagen University and then Oxford before returning to Denmark where she became an inspiring teacher. Jenny never married and in later life suffered severe psychological problems. According to her death certificate, Jenny died of 'manic depressive psychosis in its manic phase'.

In 1903, Harald and Bohr entered Copenhagen University. Both brothers played football for the university football team Akademisk Boldklub, later known as AB, which still exists in the Danish second division. Bohr played in goal, passing the time when the play was at the other end of the pitch by doing calculations on the goalpost; this lack of attention sometimes led to spectacular saves when the play suddenly returned to his half of the field. Harald was the more gifted player, representing Denmark in the 1908 Olympic Games, beating France 17–1 in the semi-finals before losing the final 2–0 to Great Britain.

It was not long before fellow students referred to both brothers as geniuses. From an early age, Bohr had been an avid reader and while an undergraduate he tried to keep up with the latest developments in physics. He was punctilious and gained a reputation for finding errors in textbooks. During their undergraduate years the brothers continued to live at home where their father regularly invited some of Denmark's most famous intellectuals to dinner. Bohr's father would take part in discussions with his guests after they had finished eating and Bohr and Harald were allowed to sit in as silent listeners; Bohr's later love of philosophical discussions was surely developed by these evenings.

In 1907, during Bohr's final degree year, he was awarded the gold medal from the Royal Danish Academy of Sciences and Letters for an essay on the surface tension of water that included detailed experimental work which he completed while studying for his final exams. He nearly missed the deadline for the prize; parts of the manuscript were written out by Harald, presumably from Bohr's notes when he didn't have time to write them out neatly himself.

The experiments involved precise measurements of the vibrations on the surface of a water jet. Each experiment was constructed and carried out by Bohr himself to produce a jet of water with a radius of less than one millimetre. Using nozzles made of glass that he had himself blown, he measured the speed of the water jets by cutting the flow twice at the same point at a given interval of time, capturing the length of the cut segment using photographs. The vibrations on the surface of the jets were also measured using photographs. To minimise external sources of vibrations, he had to carry out most of his experiments in the early hours of the morning when traffic was quiet.

In 1909 Bohr started his PhD thesis, *An Investigation into the Electron Theory of Metals*. The electron had been discovered just over a decade before at Cambridge by Rutherford's boss, English physicist J. J. Thomson (see chapter six). From Thomson's experiments it was known that the electron had a negative charge and that its mass was thousands of times less than the mass of the lightest element, hydrogen. Thomson developed what became known as the plum pudding model of the atom, in which the negatively charged electrons were embedded in a 'cake' of positive charge, distributed throughout the body of the atom to give it its overall neutral charge.

This picture was modified when it came to explaining phenomena like magnetism in metals. A metal was imagined to be a gas of electrons in a lattice of positively charged ions. In Bohr's doctoral thesis he found inadequacies in the existing electron theory of metals. The obvious way around the problem was to question the existence of electrons – but Bohr took an entirely different approach, and here we see his astonishing ability for original thought.

He argued that the electron theory of metals broke down when, for example, metals were placed in an electric field because a new type of physics was needed to describe the sub-atomic world. Although Bohr did not develop this new theory in his dissertation, his arguments were painstakingly backed up with calculations and rational arguments and in 1911 he was awarded his PhD. Winning a Carlsberg scholarship to spend a year studying abroad, he set off for Cambridge to work at the Cavendish Laboratories alongside the electron's discoverer, Thomson.

Sadly, things did not work out in Cambridge as Bohr had hoped. Thomson had little time to interact with post-doctoral researchers at his prestigious laboratory. In addition, Bohr's English was poor and Thomson quickly lost patience with Bohr's incorrect terminology and faltering speech. As an example, when discussing electricity Bohr referred to 'load' rather than 'charge'; Thomson had neither the time nor the patience to figure out what the young man meant. The low point came when Thomson interrupted Bohr and dismissed his ideas as rubbish, then proceeded to put forward the same argument using the correct terminology.

Bohr equally had a lot to learn about being tactful; keeping up his undergraduate habit of finding mistakes in textbooks, he pointed out to the great Thomson some mistakes in his own work. Even though Bohr was correct, he alienated Thomson to the point that Bohr started looking for a different placement. He wanted to stay in England and with the determination that was so characteristic of him, he set about improving his English by reading Charles Dickens with a English–Danish dictionary by his side.

In October 1911, Bohr attended the annual Cavendish dinner where the after-dinner speaker was Ernest Rutherford. By 1911, Rutherford was heading up the physics department at Manchester University. Bohr was enraptured by Rutherford's talk and afterwards, in conversation, Rutherford was impressed by Bohr's energy and vitality. He sensed in their first conversation that Bohr was a young man with incredible potential. Rutherford said, 'This young Dane is the most intelligent chap I have ever met.'

Rutherford had recently discovered the atomic nucleus that resulted in a theory of the atom very different to Thomson's 'plum pudding' (see chapter six). Rutherford's suggestion was known as the solar system model because of its similarity to our own planetary system. Here, negative electrons orbited a positively charged nucleus with the tiny nucleus at the centre of the atom containing the positive charge and nearly all the atom's mass. The trouble with this model was that, in classical physics, the accelerating electrons in an orbit about a nucleus would radiate electromagnetic waves and should spiral into the nucleus.

Rutherford invited Bohr to join him in Manchester and, disillusioned with his lot at Cambridge, Bohr accepted. He packed his bags and in 1912 headed north to work in Rutherford's burgeoning department. The two men soon established a close rapport and Bohr told Rutherford that he wished to work on the riddle of electron orbits; he wanted to know why they did not crash into the central nucleus. Rutherford considered the problem to be too difficult to solve any time soon but as he thought so highly of Bohr's abilities he let him see how far he could get.

Bohr set about finding a way that electrons could orbit without radiating away their energy. Although Bohr had an undergraduate degree in physics he was largely ignorant of spectroscopy, which was mostly confined to chemistry. Since the mid-1800s it had been known that each element had a unique spectral fingerprint. When the light given off by salts sprinkled in a flame was analysed, the bright lines which appeared in a spectroscope were different for each element. In fact, the element helium had been discovered first in the spectrum of the Sun (hence it is named after Helios, the Greek sun god) long before it was detected on Earth.

Bohr had a hunch that the unique spectrum of each element said something fundamental about the structure of atoms and searched the literature on spectroscopy. The best-known spectrum was that of hydrogen – and in 1885 a Swiss schoolteacher, Johann Balmer, had found an empirical formula which fitted the observed spectrum perfectly. Balmer's formula worked, but no one had any idea why it worked.

This is where Bohr started to apply his ability to think in an unconventional manner. After playing with the Balmer formula for a while he realised that it could be written in a different way by using the constant h, which Planck introduced for his light quanta in black-body radiation and Einstein used to explain the photoelectric effect. It seemed that the spectrum of hydrogen was somehow linked to the quantum theory of Planck and Einstein.

The next challenge was to find out what was happening. By the laws of classical physics the electron should have radiated only one frequency, as its acceleration was constant. Instead, it sometimes produced red light, sometimes green and sometimes blue. Bohr real-ised that if the electron could orbit the nucleus in different orbits then it would radiate at different frequencies depending on which orbit it was in. A larger orbit meant a higher frequency and blue light while a smaller orbit resulted in lower frequency and red light. Bohr quickly saw the link – the different lines in hydrogen's spectrum could be due to the varying electron orbits.

But what were these orbits? What determined their size? Anything orbiting has angular momentum and Bohr decided to see what happened if he only allowed the angular momentum to have certain values that related to the circumference of the orbit and Planck's constant. When he put the numbers in, everything fell into place. He was able to derive Balmer's formula from a different premise – that the electron could only orbit the nucleus in certain orbits that he dubbed 'allowed states'. Bohr had extended the existing idea – that the emission and absorption of light was quantised – to suggest that the very structures of atoms were quantised.

At the end of the summer of 1912 Bohr left Manchester to return to Denmark where he married Margarethe Norlund, a student he had met the previous summer. He was soon appointed as a junior lecturer at Copenhagen University. However, Bohr maintained close ties with Manchester, writing frequently to Rutherford about his latest ideas. He set about writing up his work on the quantisation of electron orbits but Bohr was a painfully slow writer who went through dozens of drafts before he was happy with what he had written. He sent each draft to Rutherford, who started to lose patience with the process.

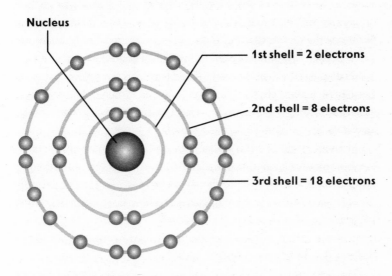

Figure 8: a schematic diagram of the Bohr model of an atom. Bohr argued that electrons could only occupy certain orbits, and later that only a maximum of two electrons could exist in any orbit. This means that the first shell can have up to two electrons, the second shell up to eight electrons (as there are four possible orbits in the second shell) etc. Not only did this model help explain the emission and absorption spectra of atoms, but it also explained the periodic table of the chemical elements.

In March 1913, Bohr sent Rutherford what he hoped was the final draft. After reading through it line by line, Rutherford wrote back, 'I suppose you have no objection to my using my judgement to cut out any matter I may consider unnecessary in your paper?' Bohr caught the first boat he could from Copenhagen and when he arrived in Manchester spent the following day and night trying to defend his paper. Bohr doggedly argued with Rutherford, justifying line by line what he had written. Eventually, Rutherford buckled under the onslaught and agreed that the paper should be submitted as it was. The paper was sent to the *Philosophical Magazine* and caused quite a stir.

Suddenly, Bohr was the wayward child of atoms with a theory that made no sense to most physicists. The idea of light quanta was still not widely accepted and physicists were hardly enthusiastic about extending the idea to the very structure of atoms. For many scientists, the quantum was a German invention cooked up by Planck and Einstein and it was widely dismissed as fantasy, even in Germany many spoke out against it. German theoretical physicist Max von Laue was so incredulous of Bohr's ideas that he said, 'If this theory is correct I shall quit physics.' Fortunately, he was dissuaded from such drastic action.

Many objected to Bohr's idea because he combined concepts of classical physics such as angular momentum with quantum theory. But Bohr was able to point to the success of his model in explaining the Balmer formula for the red, green and blue lines observed in the spectrum of hydrogen and few could argue with that. Few, that is, until more detailed observations of hydrogen spectra revealed features that Bohr's model failed to predict.

The red, green and blue lines proved not to be single lines; at higher resolution each split into several closely spaced lines. Bohr's model could not account for this. Another early supporter of the quantum theory, German theoretician Arnold Sommerfeld, applied the special theory of relativity to the orbits of electrons and found that they need not all be circular but could also be elliptical. This was able to explain the extra lines and the model was saved.

Still, the idea of quantised orbits troubled many physicists. The fundamental differences between classical physics and quantum theory seemed insurmountable. In addition, the Bohr-Sommerfeld model only correctly predicted the spectral lines of hydrogen; it didn't work for other elements. Bohr struggled with horrendously complex calculations to extend his model but was unable to make headway. Combined with the teaching demands of his job at Copenhagen, Bohr was exhausted by his work every day. When, in 1914, Rutherford offered him a research position in Manchester with no teaching duties, Bohr readily accepted.

Just before Bohr could take up his position, war broke out between Germany and Great Britain. Although Denmark remained neutral, German and British warships fought off the Danish coast in a battle that saw Britain gain control of the North Sea. This assured safe passage to neutral shipping and Bohr could board a ferry to England to move to Manchester. However, his crossing was not straightforward as he and his wife were forced on a long detour through a storm and fog around the coast of Scotland before they eventually arrived.

Despite the war, news from work in Germany filtered through to Manchester and Bohr learned that theoretical predictions made by Sommerfeld and others from Bohr's atomic theory were producing great advances. Although many questions about Bohr's model remained unanswered, the fact that it explained so much about atoms and was in such good agreement with experimental results meant that it could not be abandoned. One of the theoretical predictions coming out of Sommerfeld's group excited Bohr so much that he was determined to devise an experiment to test it. Bohr needed complex apparatus and turned to Rutherford for help in procuring this. Sadly, Bohr went on to accidentally set the equipment on fire, reducing complex, hand-blown glasswork to a heap of exploded splinters.

Bohr's return to experiments might have been brief but his theoretical work was gaining more acceptance and the Danish authorities realised that they could not lose his genius to Manchester. The University of Copenhagen offered the thirty-year-old Bohr a professorship to entice him back to his homeland, promising funds to set up a special institute of theoretical research.

So in 1916, as German and British troops were confronting each other off the coast of Jutland, Bohr and his wife made the hazardous journey back to Copenhagen. By this time Margarethe was pregnant; their first child was born a few months later in Copenhagen. Bohr named his son Christian after his father, who had died a few months before Bohr met Rutherford. Impressed by the happy atmosphere and effectiveness of Rutherford's group, Bohr modelled his institute on Rutherford's laboratory. In the years to come, the Bohr Institute would become the intellectual centre of the development of quantum theory.

Bohr was an informal person, a trait which led to an embarrassing incident when he became a professor at Copenhagen University. As a newly appointed professor, he had to present himself to the King of Denmark wearing a morning suit and white gloves. When Bohr turned up, the King shook his hand and said that it was an honour to meet such a great footballer. Bohr had the temerity to correct his monarch, pointing out that it was his brother Harald who had played for Denmark. The King was dumbstruck at such a breach of etiquette. He repeated his comment and Bohr compounded his error by this time pointing out that, although he was a footballer in his youth, it was his brother who was the great player. Outraged, the King announced, '*Audiensen er jorbi!*' (the audience is finished) and Bohr was ushered out of his sight.

In 1918, work began on the Institute for Theoretical Physics which, despite its name, was to be equipped with laboratories along the lines that Bohr had admired in Manchester. Initial funds were provided by Carlsberg, the brewers of the famous lager, but Bohr was forced to search for additional funds for the eventual completion in 1921. As soon as it opened, the institute began to attract the brightest theoretical physicists from all over Europe.

One of the successes of Bohr's atomic model was its prediction of a new element. He had further developed his theory to argue that orbits would become filled as more electrons were added to an atom, so heavier elements would have more electron shells. He argued that the chemistry of a particular element was determined solely by how many electrons it had in its outermost shell. This is standard knowledge today but Bohr was the first to suggest it. On the basis of this model, Bohr realised that there was a missing element with atomic number seventy-two, and was able to predict its chemistry based on the electrons it should have in its outer shell.

Using spectral analysis, element number seventy-two was identified at the laboratories in Bohr's institute and is called hafnium after the Latin name for Copenhagen. The prediction and discovery of this new element did a great deal to persuade theoretical physicists of the validity of Bohr's model, but the exciting discovery soon ran into difficulties. Hafnium's existence was disputed by Arthur Scott, a seventy-six-year-old Irish experimentalist who announced that he had discovered it nine years earlier and had given it the name celtium.

The dispute made the popular press, which characterised it as a question of national pride. For Bohr the stakes were high; the reputation of his new institute could be ruined by charges of cheating. Scott appeared in public with a test tube containing a sample of celtium, but the dispute rumbled on and Rutherford was brought in to settle the argument. He persuaded a reluctant Scott to send a sample of celtium to be spectroscopically analysed in Copenhagen. It was found to contain no trace of celtium, which has never been heard of since.

Later the same year, Bohr received the ultimate accolade in the form of the Nobel Prize. Planck had been awarded the prize in 1918 for work on energy quanta, and in 1921 Einstein finally received the prize (held over until 1922), as we saw in chapter seven, for explaining the photoelectric effect using Planck's light quanta. The citation for 1922's Bohr's prize read, 'For his services in the investigation of the structure of atoms and of the radiation emanating from them.'

Despite this official recognition of Bohr's work, the whole edifice of his atomic model was still on a shaky foundation. Although the model worked and was able to make experimentally verified predictions, no one had been able to properly marry the apparent contradictions between classical physics and this quantum model. In the quantum picture, radiation was emitted by atoms when they 'jumped' from a higher to a lower orbit; while they were in one particular orbit they emitted no radiation. This contradicted the classical view. The two ways of looking at the world were contradictory; both could not be true.

Bohr, however, believed that atoms could obey classical physics *and* quantum theory. He found that every quantum jump could be associated with an orbit following classical mechanics. Indeed, Bohr found that for lower frequencies (lower energies), quantum theory and classical theory produced precisely the same answer. This led to Bohr's correspondence principle, which states that at sufficiently low energies, quantum theory and classical mechanics become identical. The correspondence principle did not provide answers; it just gave an indication of where quantum theory and classical physics could agree. The inconsistencies of this ad hoc model troubled Bohr but he felt that they would eventually be ironed out.

By the early 1920s, most of the best minds in physics were working in quantum theory and Bohr's institute in Copenhagen had become a major centre for this research. The list of people working with Bohr was a who's who of quantum theory. From Switzerland came Wolfgang Pauli, from England Paul Dirac (see chapter nine), from Germany Werner Heisenberg and Erwin Schrödinger and from Russia Lev Landau. At some point, nearly everyone working in quantum theory passed through Bohr's institute.

For the best part of a decade, quantum theory itself had been flying by the seat of its pants. A piecemeal picture of incredible complexity had been put together but no one had figured out an underlying structure for this house of cards. One major problem was solved in 1924 by Wolfgang Pauli. He seized on a suggestion that electrons possess a property called spin – not a true rotation but with some similarities to a rotating planet. Pauli suggested that no two electrons could have exactly the same quantum state which meant that each orbit could only contain one electron with each of the two allowed values for spin, up and down. The inner shell, with only one orbit, could only house two electrons. The second shell had four orbits, meaning it could have up to eight electrons. Pauli called this idea the exclusion principle.

Not only did Pauli's idea explain why the electrons do not all end up in the first shell, it also explained the periodic table. For example, the next element in the table after hydrogen is helium, which is a noble gas. This means that it does not easily combine with other elements. Pauli argued that this was because with two electrons, its inner shell was full and so it did not seek any other elements with which to combine to fill its outermost shell.

The next element in the periodic table is lithium, with three electrons. Lithium is highly reactive. In Pauli's model, lithium's third electron sits on its own in the second shell, meaning it will do all it can to combine with other elements to divest itself of this lone electron. With two electrons filling the first shell and eight filling the second shell, the element with its two inner shells filled would have ten electrons. Neatly, the element with atomic number ten is neon, which like helium is an inert gas. Pauli's exclusion principle explained why helium and neon were chemically similar.

Although Bohr was not a co-author on Pauli's paper, he played a crucial role in its development, spending long hours talking to Pauli during the latter's many visits to Bohr's institute. Over the months that Pauli was struggling to come up with what would become the exclusion principle, he and Bohr corresponded frequently. Bohr had become a father figure to both Pauli and other young quantum physicists and gave them important insights. Bohr didn't always agree with the young guns, but he made sure he created an atmosphere in

Copenhagen where even the most outrageous ideas could be discussed.

The following year, another sensational step forward took place. Werner Heisenberg submitted a paper entitled *On a Quantum-Theoretical Reinterpretation of Kinematics and Mechanical Relations* to the editor of *Zeitschrift für Physik*. Despite being only twenty-three, Heisenberg was already a veteran of quantum physics. In June 1922, he attended a series of lectures given by Bohr at Göttingen University and was struck by Bohr's precision in his choice of words. Heisenberg would later say, 'Each one of [Bohr's] carefully formulated sentences revealed a long chain of underlying thoughts, of philosophical reflections, hinted at but never fully expressed.' At the end of Bohr's third lecture, Heisenberg pointed out some difficulties in a paper that Bohr had praised.

As people began to leave the lecture theatre, Bohr asked Heisenberg if he would like to go for a walk with him later that day. Going for walks to discuss physics was Bohr's preferred method of interacting with fellow physicists. They hiked some three hours to a nearby mountain and Heisenberg later wrote that 'my real scientific career only started that afternoon'. Bohr invited Heisenberg to Copenhagen for a term and Heisenberg jumped at the opportunity.

On 15 March 1924, Heisenberg found himself standing in front of the three-storey, neo-classical edifice. Once inside, he discovered that only half of the building was used for physics, with the other being given over to accommodation. Bohr and his family lived in an elegantly furnished apartment occupying the entire first floor of the building. The top floor had rooms for domestic staff and guests. When Heisenberg arrived there were six permanent staff and nearly a dozen visitors; the institute was struggling to find space.

Bohr was already making plans to extend the building, though it had only been open three years. Over the next two years neighbouring land was purchased and two new buildings added that doubled the institute's capacity. Bohr and his family moved into a purpose-built house next door and more office space, a dining room and a self-contained three-room flat was created in the original building. It was here that Pauli and Heisenberg often stayed during their many visits.

Heisenberg was only with Bohr for a month initially, but returned in September. He would later say, 'From Sommerfeld I learned optimism, in Göttingen mathematics, from Bohr physics.' For the next seven months, Heisenberg thrived in the atmosphere and saw firsthand how Bohr was trying to overcome the problems plaguing quantum physics. Heisenberg found that Bohr would talk of nothing else. Having taught Heisenberg all that he could of these problems, Bohr had great hopes that the young man would help solve them. Heisenberg's response was a tour de force in matrix mechanics, a mechanism to predict the outcomes of quantum physics that abandoned models which could be visualised to work purely with observed values.

Although matrix mechanics was a clear step forward, Bohr did not like the way it disposed of a physical model of the sub-atomic world. Also, physicists at the time were completely unfamiliar with the strange mathematics of matrices. Heisenberg himself remarked, 'I do not even know what a matrix is,' and many physicists hoped for a different explanation of the quantum world. They did not have to wait long; within a matter of months Austrian theoretical physicist Erwin Schrödinger offered an apparently completely different explanation that was much more readily accepted by the physics community.

Schrödinger formulated an approach he described as wave mechanics, producing a wave equation that described electron orbits in terms of standing waves – like waves in a skipping rope where an end is fixed. Heisenberg attended Schrödinger's talks on his ideas and, when he was next in Copenhagen, recounted his version of events to Bohr, who invited Schrödinger to the institute. When Schrödinger stepped off the train in Copenhagen, Bohr was waiting for him. After an exchange of pleasantries, battle commenced. According to Heisenberg this 'continued daily from early morning until late at night'. Bohr installed Schrödinger in the guest room at his home to maximise their time together and used every minute of every day to convince Schrödinger that his wave mechanics was wrong.

During one heated discussion, Schrödinger said, 'The whole idea of quantum jumps [is] a sheer fantasy,' to which Bohr quickly replied, 'But it does not prove that there are no quantum jumps.' All it proved, Bohr continued, was that 'we cannot imagine them'.

At one point, Schrödinger snapped, 'If all this damned quantum jumping were really here to stay, I should be sorry I ever got involved with quantum theory.'

To heal the rift, Bohr replied, 'But the rest of us are extremely grateful that you did. Your wave mechanics has contributed so much to mathematical clarity and simplicity that it represents a gigantic advance over all previous forms of quantum mechanics.'

When Schrödinger returned to Zurich he wrote a letter to Wilhelm Wien in which he said of Bohr, 'He is completely convinced that any understanding in the usual sense of the word is impossible.'

In the months following Schrödinger's visit, Bohr became preoccupied with finding an interpretation of quantum mechanics; it was all that he could talk about with Heisenberg who was, by now, based in Copenhagen as his assistant. The thing that troubled Bohr most was wave-particle duality. As Heisenberg later said, 'Like a chemist who tries to concentrate his poison more and more from some kind of solution, we tried to concentrate the poison of the paradox.'

Bohr decided he would try and understand the physics behind the competing wave and matrix mechanics approaches that were both highly mathematical. He wanted to grasp the reality behind wave-particle duality. Bohr felt that, if he could reconcile the contradictory concepts of a particle sometimes behaving like a wave and sometimes like a particle, he could solve the riddle of quantum mechanics. One of the things that troubled Bohr most was that there were two mathematical descriptions of quantum theory – but only one reality.

That particular problem was solved by English theoretician Paul Dirac (see chapter nine). Dirac visited Bohr's institute in September 1926 for a six-month stay and in the autumn showed that matrix mechanics and wave mechanics were mathematically equivalent. Now Bohr was more determined than ever to find a physical interpretation of the theory.

After months of late-night discussions with Heisenberg, he took a four-week skiing holiday in Norway. During his absence, Heisenberg came up with his idea of the uncertainty principle, exposing a fundamental difference between classical mechanics and quantum mechanics. According to this principle it is impossible to precisely determine simultaneously both the position and momentum of a particle.

After wading through the administrative business that had built up in his absence, Bohr got to the uncertainty principle paper which Heisenberg had put on his desk. After he had carefully read it, the men met and Heisenberg was stunned when Bohr told him that it was 'not quite right'. Bohr had spotted an error in the thought experiment that Heisenberg used in his paper to illustrate his idea. He went on to show Heisenberg that the uncertainty relations in his paper could be derived from using the wave model of an electron rather than Heisenberg's particle model.

Bohr tried to persuade Heisenberg not to publish. Heisenberg later recounted, 'I remember that it ended by my breaking out in tears because I just couldn't stand this pressure from Bohr.' The older man backed off and on 22 March 1927 Heisenberg posted his paper on the uncertainly principle, *On the Perceptual Content of Quantum Theoretical Kinematics and Mechanics.*

Bohr had not been idle either. While skiing in Norway, he had come up with his idea of complementarity. To Bohr this was a principle, fundamental to the very operation of quantum theory. In the principle of complementarity, Bohr felt he had found the key to the paradox of wave-particle duality. His idea was that the wave and particle properties of electrons and photons, of matter and radiation, were mutually exclusive and yet complementary aspects of the same thing. Waves and particles were two sides of the same coin; different ways of looking at the same reality, but they could not both be seen at the same time. Describing something as either a wave or a particle was incomplete, Bohr argued. For a full description of the quantum world, both had to be taken into account. When Bohr saw Heisenberg's uncertainty principle also stated that one could not know the energy of a particle at a particular moment with unlimited accuracy, he knew that the uncertainty principle supported complementarity.

Energy and momentum are properties we associate with particles, while frequency and wavelength belong with waves. Planck's equation $E=hf$ and Frenchman Louis de Broglie's formula $p=h/\lambda$ (where p is the momentum and λ is the wavelength) had shown that energy and frequency could be related and so could momentum and wavelength. Each of these equations, well known to Bohr, contained a quantity associated with particles and a quantity associated with waves. Having both a wave and a particle characteristic in each of these equations had troubled Bohr. So, as he read through Heisenberg's paper on the uncertainty principle, Bohr spotted that it arose from the ability to measure two complementary but mutually exclusive classical concepts; either particles and waves or momentum and position.

In September 1927, the International Physics Congress was convened in Como, Italy, to mark the hundredth anniversary of the death of Alessandro Volta. Bohr was due to give a paper and was finalising his notes until the day of his lecture. Amongst the audience eager to hear what Bohr had to say were German theoretician Max

Born, de Broglie, Heisenberg, Pauli, Planck and Sommerfeld. Bohr revealed complementarity for the first time and discussed Heisenberg's uncertainty principle and the role of making measurements in quantum theory. Bohr interwove each element and included Born's probabilistic interpretation of Schrödinger's wave function. His aim was to create the foundations for a new understanding of quantum mechanics, which later would be dubbed the Copenhagen interpretation.

Scientists had always assumed that when they conducted an experiment they were passive observers. There was a clear distinction between the observer and the thing being observed. The Copenhagen interpretation removed this distinction, arguing that the act of measuring led to the 'collapse of the quantum state' from an unknowable quantity into a measured quantity. In Bohr's view, reality did not exist without measurement. Before making a measurement of an electron it has no location or velocity, only probabilities. The act of measurement makes the electron 'real'. Einstein could not accept this idea and the two men argued over quantum physics for decades.

In December 1931, the Danish Academy of Sciences and Letters chose Bohr as the next occupant of the *Aeresbolig* (the House of Honour), a mansion built by the founder of the Carlsberg breweries. The company also provided an unlimited supply of lager to the occupant. By this time, Bohr had turned his attention to trying to understand the atomic nucleus and in the late 1930s put forward a theoretical model of the nucleus which helped clear up many of the conflicting ideas that had emerged from experiments in nuclear physics. He viewed the nucleus as a group of particles held together by short-range forces, similar to the molecules in a droplet of liquid.

He was also the first to describe what took place during nuclear fission, when larger atomic nuclei fragment into smaller ones. Fission was first achieved in 1939 by German physicist Otto Hahn and his colleague Lise Meitner. As a Jew, Meitner was forced to flee Nazi Germany for Sweden in the middle of their work. Hahn secretly sent her copies of their experimental results that showed nuclear fission had taken place. Meitner passed the news on to Bohr in nearby Copenhagen; he realised at once the vast amount of energy that would be released when an atomic nucleus was split. On a visit to the USA later in the year, he warned Einstein that Germany had the theoretical know-how to work on an atomic bomb.

When Denmark was overrun by Germany in 1940, Bohr did his best to maintain his integrity in the face of Nazi interference and remained in secret contact with British scientists. In 1941 he was visited by Heisenberg. Their relationship had cooled considerably by this time; Heisenberg was one of the few top physicists who chose to stay in Germany and had a leading role in their atomic bomb programme. However, during their meeting, Heisenberg passed on to Bohr a diagram that revealed how far the Nazis had got in developing the weapon. Heisenberg later claimed that his intention was to encourage scientists on both sides to abandon work on such a destructive device but Bohr interpreted it differently for reasons that were not clear, according to Bohr biographer Paul Strathern.

In September 1943 Bohr heard that his opposition to the Nazi occupation of Denmark was to result in his arrest. Bohr and his family travelled to a house in the suburbs of Copenhagen and once night had fallen they crawled across a field to a deserted beach where they were met by a fishing boat which ferried them 25 kilometres (15 miles) across the water to neutral Sweden. Bohr was rushed to Stockholm where the British arranged for an unmarked Mosquito bomber to collect him.

Bohr was put into the empty bomb bay and, under the cover of night, the aeroplane took off for Britain. It dodged the *Luftwaffe* during its flight over Nazi-occupied Norway and flew out across the North Sea. By now the fifty-seven-year-old Bohr was nearly freezing to death and by the time the plane landed safely in England he was almost unconscious, suffering from hypothermia and lack of oxygen. Bohr went on to travel to Los Alamos, USA, where he joined the top secret Manhattan Project, established by President Roosevelt to construct an atomic bomb, and played a key role in the development of the weapon.

Once the war was over, Bohr returned to his beloved institute in Copenhagen and continued to work as director of the institute, now with the help of his son Aage, who later won the Nobel Prize for his work on the liquid drop model of the atomic nucleus that Bohr had started in the late 1930s.

When Einstein died in 1955, Bohr took on the title of greatest living scientist and spent his last few years campaigning vigorously for the international sharing of research into nuclear fission. Bohr died in 1962 at the age of seventy-seven. During his long career he had not only begun the quantum revolution of the atom but had also nurtured and promoted the careers of dozens of theoretical physicists who went on to transform our understanding of nature. One of those men was the English theoretician Paul Dirac, the subject of our next chapter.

....

CHAPTER NINE

Paul Dirac

....

The tail-end, No 10 on the list, is Paul Dirac. Despite his relative lack of fame outside physics, this was the man of whom Einstein said, 'This balancing of the dizzying path between genius and madness is awful.' Dirac is now suspected by some of having been on the autistic spectrum: conversations with him were notorious for their awkward silent gaps, punctuated by remarks that were either monosyllabic or just plain strange. He often introduced his wife, for instance, as 'Wigner's sister' (referring to her brother, Hungarian physicist Eugene Wigner).

Paul Adrien Maurice Dirac claimed to have had a miserable childhood, which he mostly blamed on his father, Swiss émigré Charles Dirac. Charles moved to England in his early twenties and taught languages in a number of schools before settling in Bristol where he headed the modern languages department of the Merchant Venturers' School in 1896. It was at the city library that he met the woman who would be his wife and Paul's mother, Florence Holten, twelve years his junior. Their first son, Felix, was born in 1901 and Paul in 1902. A sister, Betty, arrived in 1906.

From an early age, the boys' father encouraged home education, speaking to them only in French while their mother spoke English (it was said that, when young, Paul believed men and women spoke different languages). Dirac would later look back on his childhood as a time dominated by a cold relationship with an emotionless father (though, oddly, one of the few items of memorabilia from this period are letters sent when Charles was at a conference; he wrote to the children in English and they replied with apparent affection).

Dirac gave as an example of the strained family relationship the fact that his parents rarely ate together – he was usually left with his father while the other children ate with their mother in the kitchen. This environment has sometimes been given as an alternative explanation for his odd social behaviour. Dirac once wrote, 'I did not know of anyone who liked anyone else – I thought it did not happen outside novels.'

At school Dirac was regarded as a dreamer who didn't reach his full potential and his communication with other children was limited. He was generally a well-behaved child, both in his early years and during secondary education at his father's school. His physics teacher realised that Dirac was so far ahead of the class that he simply sent him to the library with a list of books to read. Dirac stayed in the same premises to attend the Merchant Venturers' College, studying for an engineering degree two years early, at the age of sixteen. He seemed to find the focused attitude of the students better aligned with his own sober concentration on work.

Although following his brother Felix into engineering seemed an obvious choice, it wasn't long before it became obvious that he was a poor fit. The course was very practical while Dirac's expertise was all in the minor maths and physics aspects, leaving him once more topping up his learning in the library. By the time he reached his final year his mind was more occupied with the breaking news on Einstein's work on relativity than with electrical engineering.

Guided by his father, Dirac hoped to move on to Cambridge to study maths or physics but the costs were beyond Dirac's means, despite winning a scholarship. At nineteen he was left with a first-class degree in engineering and no place to go. Luckily, his lecturer, Ronald Hassé, got Dirac a free place at Bristol University on a mathematics course that he was allowed to complete in two years.

Dirac found the work insufficiently challenging and so he also attended physics lectures, which was where he first came across quantum mechanics.

As Dirac's course came to a close, Hassé approached Cambridge, where Dirac was able to go forward with a postgraduate placement to study relativity. He was supervised by physicist Ralph Fowler, not himself a great expert in relativity but a leading light in quantum physics. In October 1923, Dirac made his first move from Bristol, entering the imposing St John's College next door to Newton's Trinity. At the time, Dirac's strong Bristol accent made him a clear oddity among the largely public-school-educated students.

It was at St John's, particularly in the convivial dining environment of hall (reminiscent of a *Harry Potter* movie), that Dirac's lack of conversation became infamous. In a place where discussion ranged far and wide, high and low, he was rarely heard to speak. One of the longest exchanges that was reported was his response to a comment of, 'It's a bit rainy, isn't it?' Dirac is said to have walked to the door, peered out and returned to reply, 'It is not now rainy.'

Dirac had no interest in social life and societies. Just as in Bristol, he was far more likely to spend his evening in the library than the pub. His breaks were taken in the form of walks on Sunday mornings when he would head off into the countryside, try not to let mathematics cross his mind and return, refreshed, to the fray on Monday.

A wide range of influences began to shape Dirac's thinking. From Fowler's lectures he found out about Niels Bohr's quantum model of the atom and began to study the Danish physicist's work (see chapter eight). Despite technically being a mathematician, Dirac regularly attended high-flying physics clubs and seminars in Cambridge's Cavendish Laboratory, then under the leadership of Ernest Rutherford (see chapter six). It was here that he met two of his few friends, the young physicists Patrick Blackett and Peter Kapitza.

Dirac's postgraduate years were marked by steady success but no break-out achievement. He wrote a number of well-received papers and, apart from a few months when he was shaken by the news of his brother's suicide, he kept up a relentless pace of work. It was enough to gain him an extra three years' grant from the same Great Exhibition fund that had brought Rutherford to England, but he was not as yet singled out as a rising star. However, Dirac's interest in quantum physics was growing and was encouraged by Fowler. He had heard both Bohr and Werner Heisenberg speak and seemed a natural person to consult when Fowler received a proof paper from Heisenberg to review.

The paper, written in German, put forward Heisenberg's concept of matrix mechanics. This was initially applied to the behaviour of a single electron in one dimension to provide a mathematical description of the electron's behaviour. As we saw in chapter eight, unlike most physicists struggling with quantum theory, Heisenberg made no attempt to match his workings to a model of reality. Matrix mechanics simply provided a series of matrices – two-dimensional arrays of numbers – that described the electron's behaviour. This was pure number crunching, using an unfamiliar mathematics where A x B was not equal to B x A. There was no analogy, no picture to get your mind around. It was a mathematical black box.

Crucially, though, in disposing of the idea that electrons inhabited orbits around the nucleus of an atom, matrix mechanics limited itself to values that could be observed, providing probabilities of electrons jumping between two states. And it worked. For Dirac, the physics seemed unnecessarily messy, but one thing did attract him – those 'non-commutative' matrices that produced different results depending on the order they were multiplied. Heisenberg had seen this as an embarrassment but to Dirac it was the most interesting feature. Unlike most physicists he was familiar with this kind of mathematics.

The realisation came to Dirac on one of his walks that he could make something of the difference between the two multiplications, specifically when combining the representations of position and momentum to produce a mathematical formulation of the strange quantum behaviour that seemed analogous to the expected classical behaviour. Although in itself Dirac's equation was abstract, it could be used to produce values with predictions that could be tested in the real world.

Heisenberg responded warmly to reading of Dirac's discovery, though he warned that much of the content of Dirac's paper was already known in a scattered form. But Dirac had made his mark on the international community as a player. For the next few months, into 1926, Dirac continued to work on his own mathematical approaches to quantum mechanics. The field seemed very much in the ascendant despite not having made any useful predictions to show it was superior to the early quantum ideas built on Bohr's atomic model.

Part of the problem in dealing with reality was that many of the particles involved were moving quickly, meaning that Einstein's special theory of relativity (see chapter seven) should apply. As yet, quantum mechanics ignored relativistic effects. Dirac, whose first love in physics had been relativity, was aware of this but struggled to bring relativity into his work.

As the university year reached its climax in May and June 1926, Dirac was writing up his PhD thesis on quantum mechanics. The timing was difficult. News was just coming through from the Continent of an alternative approach to Heisenberg from Erwin Schrödinger. Schrödinger's approach built on something far more capable of visualisation than Heisenberg's stark matrices – a wave equation describing the behaviour of quantum particles.

Physicists liked waves; they were brought up on them. And Schrödinger's approach made it possible to explain the fixed orbits Bohr had proposed for electrons in an atom. They required a fit of complete half-wavelengths to the orbit, just as waves in a rope with a fixed end have to fit in multiples of half a wavelength, as the fixed part of the rope is forced to remain immovable.

It wasn't entirely clear yet what the waves represented, but they made quantum physics more approachable. In the end, with a deadline looming, Dirac decided to ignore Schrödinger's work and submitted his thesis using only Heisenberg's formulation. It proved a big success with the Cambridge physicists, making Dirac's doctorate a foregone conclusion.

With the formalities out of the way, Dirac could concentrate on getting on top of Schrödinger's wave approach. He managed to generalise Schrödinger's equation to deal with many of the situations that changed with time, as initially the equation had only worked in steady states. As it happened, Schrödinger had also managed to reach this conclusion and beat Dirac to publication. But Dirac was soon to make a unique contribution.

Dirac considered how the wave approach could apply to groups of quantum particles. Such particles come in two distinct types, now called fermions and bosons. Bosons, like photons of light, are gregarious – as many as you like can accumulate in the same quantum state, including position. But fermions, like electrons, obey the Pauli exclusion principle that says that no two particles can be in exactly the same state. In the electron orbits around an atom, for instance, an additional electron is forced to join a different orbit if there are already electrons in all possible states.

Dirac discovered that there is a change in the wave equation when fermions swap places but the same does not apply to bosons. This was far more significant than it sounded and shed light on Planck's formula. Planck had first established the quantum nature of light to explain the way that bodies radiated, but it had been a pragmatic fix. It worked, but no one knew why. Dirac's understanding of Schrödinger's equation as applied to exchanges of particles made it possible to derive Planck's formula from quantum mechanics. Italian physicist Enrico Fermi had produced a similar result by different means a few months before, but this didn't undermine the importance of Dirac's work.

When the 1926–7 academic year began, Dirac found himself away from his familiar haunts, first spending six months at Bohr's institute in Copenhagen. Here, Dirac was among a number of physicists working on quantum theory who came to the realisation, published by Max Born, that Schrödinger's equation (or more accurately, the square of its outcome) represented the *probability* of finding a quantum particle in a particular location. This was something of a relief, as when first proposed it was assumed that the equation dealt with a particle's location that seemed to spread out over time. Particles clearly didn't behave like that.

Dirac and Bohr got on surprisingly well, considering their stark differences. Bohr was deeply interested in culture and was someone for whom words were highly important – for Dirac, physics was all about the maths and nothing outside science mattered. Yet they enjoyed each other's company, apart from the time when Dirac was asked to help Bohr with a paper. Perhaps because his handwriting was appalling, Bohr liked others to write his words down as he repeatedly re-formulated sentences. Other junior physicists seemed happy to be Bohr's scribe but Dirac left within minutes, commenting, 'At school I was always taught not to start a sentence until I knew how to finish it.'

In his long spells of working solo in Copenhagen, Dirac realised that there was no clash between Heisenberg and Schrödinger's versions of quantum theory. Instead, he showed that they were simply different mathematical representations that could be converted one into the other. As with many of his breakthroughs in quantum physics, this was pretty much paralleled by another scientist – in this case Pascual Jordan.

It wouldn't be long, though, before Dirac could return the favour. Jordan had been thinking about taking a quantum approach to Maxwell's classical description of electromagnetism. Specifically, he was considering the common electromagnetic process involved in the production or absorption of an electron. As we saw in chapter eight, when an electron in an atom drops down to a lower orbit, for instance, its energy is given off as a photon of light, while a photon can be captured by an electron, increasing its energy.

Dirac produced a quantum theoretical description of what was happening in the creation and destruction of photons as interactions with an electromagnetic field. A photon is then regarded as simply a 'blip' in the field – a quantum theory for radiation. He spent Christmas of 1926 with the Bohr family – his first time away from home – finding a real contrast between the cheerful environment and his own formalised family life.

In the new year, he moved on to Göttingen, where he spent the next six months in Heisenberg's home territory. Here, under the wing of Max Born, was Dirac's mental sparring partner Pascual Jordan and young American physicist Robert Oppenheimer.

In some ways, Dirac preferred Göttingen to Copenhagen – notably the beautiful countryside for his Sunday walks – but intellectually, the more restrained academic atmosphere seemed to be less effective for Dirac than the high-pressure environment of Bohr's institute. Although he developed a quantum mechanical description of scattering – the process, for instance, by which light from the sun is distributed through the atmosphere to turn the sky blue – he took no major steps forward while there.

Unusually, Dirac decided that he needed a break – a summer away from the quantum pressure before he returned, perhaps to take on one of the oddities of the theory. This was the concept of quantum particles having a quantised property called spin that clearly bore no resemblance to our usual concept of something spinning on its axis. In reality, his break seemed to consist of spending the summer in his bedroom, back for the first time in nearly a year in his parents' house, working still, but in a less conducive atmosphere.

Nothing much seems to have come out of that summer – certainly no reconciliation with his father – and it was only when he was back in Cambridge in October 1927, now a fellow of St John's, that Dirac was really able to make progress on bringing relativity properly into the quantum sphere. This time he combined the quantum mechanical description of the behaviour of an electron with the special theory of relativity to make the theory compatible with fast-moving particles and different frames of reference.

Working mostly in his sparsely furnished rooms in college, Dirac laboured tirelessly, trying to formulate an equation through some kind of happy inspiration as it was impossible to derive it directly. As the college changed in atmosphere with the end of the undergraduate term and the mounting preparations for Christmas, something emerged in a process that Dirac never adequately described. The result was a four-part equation that not only dealt with the pressing problem of spin but predicted other aspects of an electron's behaviour exactly as observed – and that collapsed to the non-relativistic predictions of existing quantum theory when an electron was moving slowly. If physicists were asked to rank their top ten equations, this would be there (but that's another book).

Despite the constant theoretician's fear of being beaten to publication, Dirac seemed in no hurry to share his equation. He mentioned it in passing to another young physicist, Charles Darwin (named after his better-known grandfather), before heading back to Bristol for Christmas, but his paper would not be sent off to the Royal Society until the beginning of 1928. The response, when it was published in February, was shock from the physics world, both at the heavy duty mathematics involved and the fact that Dirac had cracked so difficult a problem.

However, all was not rosy. While everyone admired the equation's power, it was also downright bizarre. The equation showed that an electron could not only have positive energy levels, it could also have *negative* energy. The equation seemed to suggest that an electron could make a succession of quantum leaps far down below zero energy.

The physics world was not unfamiliar with ignoring uncomfortable outcomes when an equation was useful. Maxwell's equations, for instance, predicted both the conventional electromagnetic wave of light and an inverted version that travelled backwards in time from receiver to source. That had been successfully ignored and, if you did the same with Dirac's negative energy, his equation was a marvel too.

After a summer on the continent that included a first trip to Russia, Dirac returned to Cambridge for the start of the 1928–9 academic year. Although a man of few words, he enjoyed spending time with other physicists provided he was not required to speak much and was refreshed by his summer tour. Yet his revived efforts were not applied to his equation; much of the next year was diverted into writing a textbook. He had had no insights on that difficult negative energy by March 1930, when he set off for an extended visit to the USA.

Travelling across the country, mostly by train, Dirac went from New York to Princeton and Chicago before making for Madison, Wisconsin. It was here that Dirac's curt directness was reported most effectively. At a lecture in Madison, when it came to time for questions, a member of the audience said, 'I don't understand the equation in the top right-hand corner of the blackboard.'

Dirac said nothing, resulting in an uncomfortable pause, until he was prompted and replied, 'That was not a question, it was a comment.'

From Wisconsin, Dirac continued his journey, mixing academic visits with walking opportunities in striking locations from the Grand Canyon to Yosemite. Towards the end of the trip, in August, Heisenberg joined him and after lecturing in California the pair headed off on a steamer for Japan where they were treated as celebrities. Here they split, Dirac taking the Trans-Siberian express to Moscow and flying to Berlin (air travel was still a novelty then) before finally making it back home.

Returning, probably with some relief, to his semi-monastic existence, Dirac set his sights on the negative energy problem. Earlier in
the year it had been shown that the negative energy levels could not
after all just be ignored as an oddity but were required if Dirac's equation were to be used. They were make or break for his theory. Finally
he came to a solution; but it involved a scenario that raised eyebrows
even among quantum theorists.

Dirac realised that there was a circumstance in which negative
energy states could exist but not be observed. He imagined that the
universe contained an infinite sea of electrons, filling up all the
negative energy levels before any 'real' electrons entered the fray.
Then it was inevitable that the observed electrons would have positive energy because there were no empty negative levels for them to
drop into.

The idea seemed ridiculous but if it was like any decent theory, it
should leave its mark and be testable. Such a test was possible
because the theory suggested that an electron from the negative
energy sea would occasionally be boosted out, leaving a hole in the
negative energy foundations of the universe. A normal electron
approaching would drop into that hole and as it did so it would give
off electromagnetic radiation – photons.

Experimenters would have to have a picture of this negative
energy hole to observe the process. What such a hole amounted to
was the *absence* of a negatively charged, negative energy electron.
This should be identical in behaviour to the *presence* of a positively
charged, positive energy particle – like an ordinary electron but with
a positive charge, a bit like a proton. And initially this was what Dirac
assumed such a hole must be, though the theory didn't explain why a
proton was so much heavier than an electron.

It soon became obvious that the proton interpretation of holes
would not do. It meant that atoms would be unstable, collapsing as
their electrons dropped into their negative energy protons. As 1929
slid into 1930, Dirac was faced with a theory that worked wonderfully

well, yet with a mysterious foundation. There was one highlight of those dark months, though. In February, Dirac was elected a fellow of the Royal Society, one of the society's youngest ever fellows.

Perhaps this honour improved Dirac's self-worth as, soon after, using his newly purchased car, he began to take weekdays off to relax and head into the countryside for walks rather than restricting his leisure time to Sundays. This may have made his thinking more effective but as yet it did not untangle the problem of the negative energy holes. After a short summer break in Russia, Dirac was back to his contemplation of this irritating, illusive component of the universe. Yet again, though, there was a compensation. His textbook *The Principles of Quantum Mechanics* was published.

Though not the most accessible book, free of diagrams and spare of wording, it closely followed the format of Dirac's lectures and immediately became the definitive text (for those who could get through it). Dirac's peers were enthusiastic – even Einstein praised it. But into 1931 there was no light to be shed on the model that Dirac had absolute faith in, other than to confirm that its holes were definitively not protons.

To take a break, Dirac looked elsewhere. Going back to the fundamentals of electromagnetism, he looked at the implications of the way in which an electrical charge only comes in fixed units – the size of the charge on an electron or a proton. By combining quantum theory and classical electromagnetism he was able to show that magnetic 'charge' should also be quantised and in theory could exist in isolation – a 'monopole' of north or south, though none had been observed. He also discovered that because the attraction between magnetic monopoles would be much greater than electrical charges, his theory provided a neat explanation for magnetic poles being paired.

The most surprising result (and one reason that, ever since, magnetic monopoles have been high on the list of potential experimental discoveries) was that if magnetic monopoles did exist – if just one could be discovered – it would force electrical charge to be quantised in its units, a value that otherwise has no explanation.

Monopoles are still to be confirmed, but the thinking opened up Dirac's mind on negative energy holes. At this point, there were only two known fundamental particles – the electron and the proton (if you didn't count a photon of light). But if monopoles existed, why not other particles? For instance, a particle that could accurately represent a hole, a positively charged, doppelgänger of an electron?

Dirac wrote, 'A hole, if there were one, would be a new kind of particle, unknown to experimental physics, having the same mass and the opposite charge to an electron. We may call such a particle an anti-electron.' He predicted that anti-electrons would not be easy to spot, as they would naturally tend to be eliminated by combining with a conventional electron – a process now described as annihilation – but that it might be possible to produce the anti-particles experimentally. He also suggested there should be an anti-proton.

Experimental evidence evaded any physicist who took Dirac's idea seriously. But as 1931 drew to a close American physicist Robert Millikan arrived in Cambridge to give a seminar on his pet subject, cosmic rays, featuring photographs of particle tracks taken by his PhD student, Carl Anderson. These 'rays' are actually high-energy particles from deep space which come crashing into the Earth's atmosphere, which provides a kind of natural particle accelerator far more energetic than anything available in laboratories, even today.

Anderson had noted that high-energy collisions frequently produce an electron and a corresponding, positively charged particle. To see the tracks of these particles, they were passed through a cloud chamber, a device that is supersaturated with water vapour – any passing particle forces water droplets to form to provide a visible trail. At Millikan's suggestion, Anderson set up a strong magnetic field that would mean electrically charged particles would head in a curve with the direction reflecting whether they were positive or negative. The deflection would provide a measure of the particle's momentum.

In a good few of these photos, electrons were paired with a positively charged particle that came into being at the same point in time. While this was happening, Dirac was in Princeton on a sabbatical so he missed the first experimental evidence of what would prove to be his anti-electron – or positron, as it became known. And this wasn't the only new particle on the cards.

At the start of 1932, James Chadwick, working at the Cavendish, discovered evidence for a particle that Rutherford had hypothesised some time before. This was a neutral particle with similar mass to the proton – a neutron (see chapter six). By spring 1932 the Cavendish had another breakthrough – John Cockroft and Ernest Walton split the atom by bombarding lithium with fast protons.

The experimentalists seemed to be forging ahead, leaving Dirac playing catch-up. But once again, academic powers that be signalled their belief in him. When seventy-five-year-old Lucasian professor Joseph Larmor retired, Dirac was announced as his replacement. Isaac Newton had reached the position at twenty-nine and Dirac was similarly aged. That age factor was no trivial matter to Dirac. He was aware that most great physicists did their best work before reaching thirty. He had joked to Heisenberg when he reached the milestone that he was 'no longer a physicist'.

That summer, Dirac travelled to Russia again, holidaying on the Crimean coast. While he was there, Anderson finally produced a cloud chamber image of one of the positive particles produced by cosmic rays of sufficient quality to follow its track. He discovered that it behaved as if it were a positively charged electron and was so surprised that he checked to make sure that no one had tampered with his electromagnet to reverse its polarity. Within a month he had found another two 'positive electron' tracks.

His immediate response was not to vindicate Dirac's theory,

something which didn't even occur to Anderson, but to dismiss the experimental results. It took an observant mathematician, Rudolph Langer, to make the first suggestion that this was Dirac's anti-electron in the wild – and this was promptly ignored. Even when a new cloud chamber facility at the Cavendish went into action in the autumn of 1932, the results were passed over.

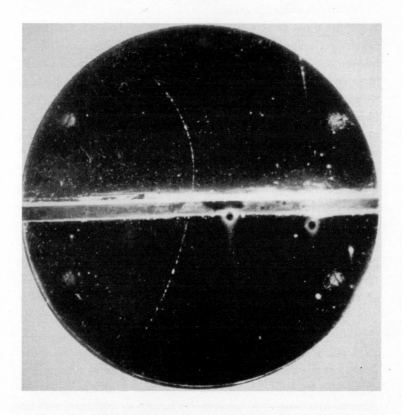

Figure 9: a cloud chamber photograph by Carl Anderson of the first positron ever identified, taken in 1932. The positron is the long curved track in the photograph, and is seen to pass through a 6mm lead plate which lies across the middle of the picture.

The problem with cloud chambers had been that almost all images taken were blank, until Cavendish physicist Patrick Blackett and co-worker Giuseppe Occhialini put Geiger counters either side of the chamber. These triggered a photo as a cosmic ray passed through and increased the efficiency of the process. They too detected the positive electron and described the result in a seminar attended by Dirac . . . who said nothing.

At the time, Dirac was working on the application of a classic physics technique that considered different possible paths for a moving object and its action – which is the sum of the differences between potential and kinetic energy over the path. As we will see in chapter ten, this is the approach that proved inspirational for Richard Feynman in understanding the quantum behaviour of electrons. But in an attempt to show solidarity with the Soviet Union, whose politics he admired, Dirac published in a new Soviet journal, meaning his idea pretty much disappeared for years.

By the start of 1933 there was too much experimental evidence to ignore the anti-electron. Dirac made calculations, assuming that the energy of collision of an incoming cosmic ray particle with atmospheric gas was being converted into a pair of particles, a clear example of the relationship between mass and energy that emerged from the special theory of relativity (see chapter seven). It took another year for the physics community to be convinced – and even then some were grudging about acknowledging that Dirac's theory truly predicted the existence of the positron. But there was no going back.

At the end of the year, Dirac received the final vindication of his position in scientific history, a call telling him that he would share the 1933 Nobel Prize in physics with Schrödinger (Heisenberg got the deferred 1932 prize all to himself). On his trip to Stockholm with his mother as his paid-for guest, Dirac said little but his mother more than compensated, taking the opportunity to criticise Dirac's father for being a tyrant and emphasising that her son was 'not interested in young women'. This didn't help a growing speculation about Dirac's sexuality.

In his Nobel lecture, Dirac made the suggestion that there might be as much antimatter as matter in the universe and it was just an accident of circumstances that our solar system seemed to be entirely pure matter. He was riding high. However, the position on the top of the pile in theoretical physics is notoriously unstable. Shortly after the award, Oppenheimer, now based in California, and his team came up with a quantum field theory that kept the positron, but disposed of the negative energy sea. Another development from Pauli showed that hypothetical spin zero particles, which according to Dirac shouldn't have antiparticles, would indeed have them if they existed. The cracks were showing in Dirac's theory.

To make matters worse, the infinities that regularly cropped up when using the quantum field theory that Dirac pioneered had not gone away, as it had at first been assumed they would, but continued to plague the models. Dirac felt that he had wasted two years and so for the first time worked with another physicist, his experimentalist friend Peter Kapitza. Their initial collaboration produced no great fruit, but Dirac was able to suggest an ingenious way to separate different isotopes of an element by forcing them into a helical path. Dirac was fascinated to discover that the exhaust from his high pressure whirlwind of gas separated itself into two streams around a hundred degrees different in temperature. The man who had been dismissive of experimentation was coming round to its virtues.

Fairly soon, another Dirac failing was to disappear. On sabbatical in Princeton, Dirac met a fellow traveller, Hungarian quantum physicist Eugene Wigner. Joining Wigner for lunch, Dirac was introduced to his friend's sister Margit, or Manci, as she was known. She was very much the opposite of Dirac – outgoing, loud, fun to be around. The kind of extrovert he always got on with. And it didn't take long to become obvious that these opposites were attracted.

While in Princeton, Dirac focused on updating his textbook and trying different ways to frame his equation for the electron. As his relationship with Manci deepened, he was forced into taking action in a very different world when he heard that his friend Peter Kapitza, a Russian citizen, had not been allowed to leave the USSR after his latest visit, following the defection of big name Russian scientists like George Gamow. Dirac and Rutherford in Cambridge led campaigns to get Kapitza returned to rejoin his family.

Manci returned to Hungary and their blossoming romance suffered from Dirac's inability to be anything but blunt in his letters. At one point, after Manci complained of his terseness, he produced a table of questions in her letters he had not answered, giving a pithy response to each'. It was not the ideal way to conduct a romance.

When Dirac's sabbatical ended in June 1935, he headed to Russia to support Kapitza, trying to encourage the authorities to allow his friend's return. But despite Dirac's efforts and those of other envoys sent by Rutherford, Kapitza, while treated well, was not allowed to leave Russia. Dirac spent nine days with Manci, making the trip seem slightly less futile.

As the new term started back in the UK, Dirac was making little progress on his work. His few friends were deserting him for new challenges at universities around the world. And though, in principle, he had Manci, he was still uncomfortable with anything that disrupted his routine. When one evening she rang him at bedtime, Dirac was angry rather than delighted,because she was intruding on his privacy. Yet despite their communication problems, the long-distance romance grew as Dirac continued with his routine.

The only clear break came when he had to leave a climbing holiday in June 1936 to return to Bristol where his father had fallen ill. Charles Dirac died before his son could arrive, triggering complex emotions. Dirac returned to the holiday shortly after and managed to climb the difficult Mount Elbrus. His tour was completed with visits to see Kapitza, Manci and Bohr – probably the three most important people in his life.

Back in Cambridge, Dirac switched his attention to the opposite end of the scale, moving from quantum theory to cosmology (a topic that ironically would become strongly entwined with quantum physics). Manci passed through the UK on the way to visit her brother in America and Dirac finally decided the time was right. On the drive up to London, he asked her to marry him.

What seemed a perfect step was soon thrown into confusion. Dirac's mother had a few minutes alone in London with Manci. Soon after, his mother wrote to Dirac, telling her son that she had received a surprising response to a question. Manci said that when married, 'I cannot allow Dirac to come into my bedroom.' Why then did she want to marry him? 'I like him very much and I want a home.'

Dirac was disturbed by his mother's letter. Could this possibly work? It seemed the heart won in a man usually so driven by the head: he and Manci were married a few days later.

Finally, now, Dirac seemed able to express his feelings in letters to Manci when he returned to Cambridge to find them a home while she was preparing for the move back in Budapest. The feelings seem to have been returned – whether Dirac's mother had misunderstood or misled in a last attempt to keep her son to herself or Manci had been teasing, their marriage had every sign of being happy. Manci, with two children from her previous marriage, settled into life as the kind of don's wife who brought life and energy to an otherwise stultifying environment.

In work, Dirac did not have any earth-shattering cosmological breakthroughs, but brought to the fore a topic that would be a favourite of popular science books (a genre Dirac detested). This was the way in which the values of a handful of key numbers have a huge impact on the nature of the universe and specifically how some of these very large numbers (like the number of protons in the observable universe) might be in some way linked. He was roundly criticised for metaphysical mumbo-jumbo.

Dirac was stung and dropped his new interest. Soon after, Rutherford died, and the new head of the Cavendish moved its research in different directions. Quantum field theory, meanwhile, had become bogged down. Dirac hoped to chip away at one of the infinities that beset the theory, the self-energy of the electron. The electron was a dimensionless point particle for which the strength of the electrical field goes up with the square of the distance away. As that distance approaches zero, the electrical field should shoot up to infinity. Dirac wondered if there was a fault in Maxwell's theory of electromagnetism on which he had based his work. Despite working on this for months, he got nowhere.

As World War II loomed, Dirac made little progress while Manci began to express a dislike for the stiff atmosphere of Cambridge. But with a child soon on the way, life went on. Their daughter, Mary, was born in February 1940. Despite near misses in the relatively light bombing of Cambridge and the much heavier attacks on Bristol, Dirac's family were unharmed by the war, though sadly his mother died of a stroke in late 1941. Dirac had far less input to the war itself than most leading physicists, turning down a request to work on cryptography, although later in 1940 he was consulted about his experiments on separating isotopes. If practical, his work would be hugely useful in the race to build an atomic weapon. The rare uranium-235 isotope needed to be separated from the much more common 238.

In 1941, Dirac put together a general theory that covered all mechanisms that might allow an isotope mix to be separated by varying one of the isotope's concentrations, for example by temperature gradients or centrifuging. The following year, Dirac was asked to join the code-breakers at Bletchley Park but Manci was pregnant again and he was reluctant to leave Cambridge. He did, however, become more involved in the British atomic bomb project; once more working on the separation of isotopes and coming up with the centrifugal technique that eventually became an industry standard. This wasn't picked up at the time, but of more immediate use was the work Dirac did between 1942 and 1943 to predict the behaviour of chain reactions in a lump of uranium-235.

Being Dirac, though, he did not undertake this work in the frenzied team environment of the Manhattan Project that produced the atomic bomb, but worked alone in his study in Cambridge. By now the Dirac family numbered six, including his two stepchildren, the couple's first daughter Mary and Florence (named after his mother, but known by her second name, Monica), born in September 1942.

Once more Dirac turned down a request to leave Cambridge to undertake war work. Even more strikingly, in late 1943, when the British and American bomb projects merged and the British scientists joined the Manhattan Project, Dirac declined to move to Los Alamos in the USA. Dirac's motivation seems mixed. He had travelled abroad before, but that was before he had a family and, after the coldness of his own home life, he was determined to ensure his children had a proper family around them. He also detested teamwork and, probably rightly, felt he would not fit in at Los Alamos. Finally, he seems to have had increasing doubts about the moral validity of the project and stopped all involvement in 1944.

The most significant contribution Dirac made to physics towards the end of the war was his revision of his quantum mechanics bible. In the process he introduced notation he had been using for several years that is still central to quantum theory. It shows a complimentary pair of quantum states as something like <A | B> which is known collectively as a bracket. The first part is known as a 'bra' and the second as a 'ket'. Allegedly, when new words were being discussed at high table in St John's, Dirac made the contribution, 'I invented the "bra",' to much consternation.

As the war ended, Cambridge pulled itself back to normal and Dirac fell into his comfortable groove, but physics seemed to be moving away from him. Over the next few years, the focus of the quantum world moved to the USA where Richard Feynman (see chapter ten), Julian Schwinger and the UK's Freeman Dyson pulled together a theory (simultaneously developed by Japanese physicist

Sin-Itiro Tomonaga) called quantum electrodynamics (QED) to describe the interaction of light and matter. Dirac commented, 'I might have thought that the new ideas were correct if they had not been so ugly.' He was thinking particularly of 'renormalisation', the fix by which QED rid itself of embarrassing infinities by replacing the runaway numbers with observed values.

The new generation was taking over and Dirac would make few further major contributions. During his next sabbatical at the Institute for Advanced Study in Princeton in 1947–8, he did push forward his magnetic monopole theory and tinkered with quantum field theory, but there was never another 'Dirac's equation.'

As with Einstein in his later years, Dirac made some interesting attempts at looking at physics problems in new ways, at one point re-introducing the ether – but his ideas, though creative, came to nothing. Unlike Einstein, Dirac faded from public consciousness. In part this was his own doing. He was persuaded to accept his Nobel Prize, told that he would get less publicity by accepting it than he would if he refused. In 1953 he refused a knighthood, to his wife's irritation.

Towards the end of his working life, Dirac spent less time in Cambridge as he disliked the new department of applied mathematics and theoretical physics and, most of all, regretted the loss of his favourite parking space. He spent far more time at the Institute for Advanced Study in the last few years that he was nominally at Cambridge.

After a terrible 1968, when Dirac's stepdaughter Judy's car was found abandoned and she went missing, never to be found, Dirac left his role as Lucasian professor in September 1969 and in 1971 accepted a post as visiting eminent professor at Florida State University, allowing Manci to escape the Cambridge weather and the stilted social atmosphere she detested.

Dirac did not leave his beloved physics but continued to lecture and throw new ideas into the mix, even if these would lead to no significant changes. In 1973 he returned to the UK briefly to accept the Order of Merit, a more prestigious award than a knighthood, but one that did not come with a title (to the end, he preferred to be simply 'Mr Dirac'). Dirac died on 20 October 1984, aged eighty-two.

Although the equation Dirac produced made the essential contribution of bringing relativity into our understanding of the electron's behaviour, he could never reconcile himself to the renormalisations of QED. It was our final physicist, Richard Feynman, one of the architects of QED, who would carry forward Dirac's work.

....

CHAPTER TEN

Richard Feynman

The holder of the No 8 position on the list, Richard Phillips Feynman, was described by the *New York Times* as 'arguably the most brilliant, iconoclastic and influential of the post-war generation of theoretical physicists.' He is the most recent physicist on the list, the only one to win a Nobel Prize after World War II.

Talk to physicists about Feynman and the chances are that smiles will spread on their faces. Not only would they regard Feynman as ingenious, they would highlight him as their favourite physicist. Unlike most top-ranking scientists, Feynman was a brilliant teacher. An internet search for video clips of him will quickly show how gifted he was at conveying the mystery and awe of the subject which had obsessed him since childhood. He had an ability to make complex ideas seem simple and made even more attractive by an accent that made it sound as if Tony Curtis was doing the explaining.

Relatively few current physicists have had the chance to attend a Feynman lecture, but many have encountered his scalpel-like dissection of physics in the collected notes called *The Feynman Lectures on Physics*, usually known as 'the red books' after their original covers. And then there are the stories. A psychologist would probably ascribe at least some of the hero-worship that Feynman induces in young physicists to his atypical extrovert character and his over-the-top exploits – from safe cracking to bongo playing.

Some were probably even true. There is no doubt that Feynman, who spent a lot of time moaning about the way stories in the history of science were sanitised, was good at embellishing his own tales. As the physicist Murray Gell-Mann put it, 'He surrounded himself with a cloud of myth and he spent a great deal of time and energy generating anecdotes about himself.' So it is entirely possible than some of the account of Feynman's life that follows emerged from his fertile imagination.

It all began back in 1918 – far in the past now, though Feynman will always seem a contemporary physicist. His parents Melville and Lucille were delighted to have a son, born on 11 May, and Melville particularly went out of his way to impress on the young Richard an approach reminiscent of the Royal Society's motto *nullis in verba*, roughly translated as 'take nobody's word for it'. Melville encouraged Richard to look past conventional labels to try to understand the true underlying nature of the things he observed.

Whether inspired by this or simple curiosity, the young Richard (known as Ritty) began his scientific quest with an exploration of radio. The radio receivers of the day, packed full of valves (vacuum tubes) were far easier to take apart and understand than a modern printed circuit. Feynman exploited this to the full, playing around with the kind of high-voltage rigs that would be out of bounds for any modern child. At the time he lived in Far Rockaway, a small district of the New York borough of Queens protruding from Long Island, a place with ample opportunities to explore the natural and man-made world.

It's arguable that the freedom he was allowed, and his father's early encouragement to take an interest in everything around him, inspired Feynman to have a much broader scope than the narrow specialities of most modern scientists. When, for example, he was in graduate school Feynman attended undergraduate biology classes for the fun of it. Giving a presentation on the nervous system of a cat, Feynman began outlining the names of different muscles. The other students pointed out that they knew those names by heart. Feynman commented that if they had learned all these names, 'No wonder I can catch up with you so fast after you've had four years of biology.'

This was at Princeton. Feynman had taken his undergraduate degree at the Massachusetts Institute of Technology (MIT). His choice had been between MIT and Columbia, but this was 1935 and the latter, an elite Ivy League institution, had already filled its quota of Jews. (Feynman was an atheist, but ethnically Jewish.) He started in mathematics at MIT because it had been his favourite subject at school where he triumphed at the obscure problems set in the local algebra league. But Feynman found university maths too impractical and switched to engineering which proved too simple and at length settled on the Goldilocks option of physics.

For all Feynman's brashness, he had a parochial mentality and was reluctant to move away from MIT once he had graduated, but the head of physics pushed him towards Princeton. Another Ivy League university, modelled on Oxbridge, Princeton had concerns about Feynman, both because of his terrible assessments outside of physics and maths (reminiscent of Einstein's problems getting into Zurich's *Eidgenössische Technische Hochschule*) and because of doubts about his ethnicity. The head of physics at Princeton asked, 'Is Feynman Jewish? We have no definite role against Jews but have to keep their proportion in our department reasonably small because of the difficulty of placing them.' At the time, Jewish students would have lived in separate fraternities. Feynman was admitted after assurances from MIT that he neither looked nor acted like a Jew.

Once he settled in at Princeton, which had an excellent physics department, Feynman hunted for a suitable PhD topic and settled on quantum mechanics. It soon seemed a mistake. He was uncomfortable with the complex mathematical nature of this science of the very small and the more he immersed himself in the topic, the less sure he was about finding a way forward.

A major problem was the nature of electrons. Still a relatively new concept in the 1930s, these negatively charged particles were dimensionless points, as we discovered in chapter nine, meaning that the mathematics describing them went to pieces. As the radius of the electron headed to zero, various characteristics – including the self-energy, the result of the electron's charge acting on itself – should have become infinite. As Paul Dirac put it at the time in his *Principles of Quantum Mechanics*, 'It seems that some essentially new physical ideas are here needed.'

Feynman wondered if this could be approached with as stark a pronouncement as Niels Bohr (see chapter eight) had used in forcing electrons in an atom to stick to fixed orbits. What if an electron simply was not allowed to interact with itself? This didn't seem enough and a second possibility occurred, shared with Feynman's professor, John Wheeler. Maxwell's equations for electromagnetism allowed for both 'advanced' and 'retarded' light, where retarded waves correspond to the ordinary photons we observe but advanced waves are photons travelling backwards in time from receiver to transmitter. Feynman and Wheeler proposed that these advanced waves existed and cancelled out the infinity-producing self-action of the electron.

This was, of course, highly speculative and still left a formidable challenge in getting an overall picture of what was happening in the interaction between matter and light. The hint of which way to go would, ironically, come from a paper by Dirac that gave Feynman the insight to make a leap of intuition.

Feynman devised a very visual representation of quantum particles. He started from the concept of a world line. This is a plot of a particle's position against time. One dimension on the chart is time, the other position in space. (The three space dimensions are condensed into one for easier drawing.) Such world lines were valuable in exploring relativity.

Feynman imagined the impossible. He knew that, unlike such objects as a ball, a quantum particle did not have a well-defined path in getting from A to B but instead could be considered as taking every possible path, each with a different (and often near-zero) probability. He imagined drawing every possible world line – and the result would be to fully describe a particle's behaviour.

Although it clearly wasn't possible to physically do this, there were well established mechanisms in calculus to produce a finite result from an infinite set of decreasingly small values – and he also knew that many of the world lines would cancel each other out or be so unlikely that they could be ignored. Although the details weren't yet fixed, it seems as if his PhD thesis was heading towards a success.

When the USA entered World War II, Feynman was asked to join a new elite team of scientists working on a bomb to make use of a nuclear chain reaction. His immediate reaction was to say 'No'. Like Einstein, Feynman had no love for the military. But the thought of Germany developing such a bomb first – all too possible given that the first nuclear fission reaction had been done there (see chapter eight) and so many of the key physicists were German – was enough to spur him into accepting the offer.

Initially, Feynman's involvement was similar to Dirac's considera-tion of separating uranium-235. Techniques had to be designed, tested and taken to production to enable the small difference in atomic weight to differentiate the reactive isotope and its more stable cousin, U-238. This was a crucial first step, but Feynman's involve-ment was not intense and he managed to get time off to finish his PhD and receive his doctorate in June 1942.

It might seem odd that, under the pressures of war, the young physicist found time to tick an academic box, but he didn't just want his PhD, he needed it. His graduate student grant was contingent on being unmarried – but Feynman was engaged to Arline Greenbaum, and he had every reason to want to be married as soon as possible. Arline was seriously ill with incurable tuberculosis of the lymph glands (her condition made worse because she had originally been diagnosed with Hodgkin's disease). By the time Feynman got his doctorate, Arline was only expected to live another year or two.

Feynman was under pressure from his parents. They thought he was too young to marry, that seeing so much of Arline put his own health at risk and that the end result could only be tragedy for their boy. But Feynman had no intention of giving way. Arline was moved to a hospital near Princeton and taken in Feynman's DIY ambulance to be married. Their relationship could never be physical but Feynman ensured Arline had the closest thing possible to a normal wedding.

By the end of the year, Dr Feynman was on the move. Recognising the difficulty of separating U-235, the US's effort had competing teams working on different methods. Feynman was involved in the most sophisticated, using a varying electrical field to try to split a beam of uranium atoms into groups by weight. It worked, but slowly.

By now it had become clear that simpler methods involving diffusion of gaseous uranium through meshes of tiny holes were more effective. Feynman's talents were being wasted and he was asked to move to Los Alamos, a former ranch school way south in New Mexico, to join the Manhattan Project proper, the vast team working on building the bomb. The name originated from the location of the Army Corps of Engineers' headquarters and the project would always be an uncomfortable interaction between the military and the more free and easy university academics.

Feynman could not commute back to Princeton to see Arline. He made it a condition of his joining the project that his wife was found a bed in a nearby hospital. Unfortunately, Los Alamos had been chosen for its remoteness. The nearest appropriate hospital was in Albuquerque, over 95 kilometres' (60 miles') drive from the project site. It would have to do.

Feynman immediately proved to be a jack-of-all-trades, valuable in an establishment where many of the workers were academics used to a very tight focus of attention. He contributed to theory, but also proved adept at repairing the temperamental electro-mechanical calculators that were employed in crunching vast reams of numbers to establish what would and wouldn't work in the construction of an atomic bomb.

Feynman was soon put in charge of the theoretical computations group making use of the calculators. Hans Bethe, the head of the theory division, quickly discovered that Feynman made an ideal sounding board for trying out new ideas. Unlike many other young members of staff, Feynman would question and criticise Bethe's suggestions. As far as Bethe was concerned, this independence and ability to cut through to the essentials made Feynman ideal for the post, even though he was junior to many of those working on the calculations.

There is no doubt that Feynman applied himself thoroughly to his work – but even in the fevered attempt to succeed in producing a working nuclear weapon before the Germans there was spare time. No one could work 24/7. Feynman's favourite relaxation activity was taking on the Los Alamos security regime. Much of the security was going through the motions rather than being effective and Feynman, with his total disregard for authority, was determined to find loopholes and exploit them, showing up any weaknesses.

He enjoyed sneaking out through a hole in the fence and re-entering through the main gate, causing confusion for the staff guarding the base as multiple Feynmans were noted on site. But his tour de force came with the secure storage of secret documents. These were held in ordinary filing cabinets, secured by padlocks. Feynman thought that such security was laughable and demonstrated this by picking the locks and tilting a cabinet when he discovered this left the bottom drawer unlocked.

When he needed to access information, rather than ask the keyholder's permission, he would extract the document, re-lock the cabinet, use the information and then return the document, much to its owner's embarrassment. The military on the Manhattan Project were not famous for their sense of humour and Feynman's constant ability to access the supposedly secure cabinets touched a nerve. They replaced the cabinets with secure units, each accessed via a combination lock requiring three double-digit numbers to be entered before the cabinet could be opened.

At first sight, Feynman appeared to be foiled. His lock-picking was useless and, despite what is often depicted at the movies, it wasn't possible to hear the tumblers falling into place. But hours of playing with the locks brought him two hacks that would make it possible to break in. First, he discovered that the locks weren't sensitive to the nearest digit. Although, in principle, there were a hundred values for each of the three numbers in the combination, the locks responded to the nearest five digits – leaving only twenty values.

This still left 20 x 20 x 20 – eight thousand – possible combinations, which was too tedious to work through systematically, except Feynman had also discovered a mechanical flaw in the lock mechanism. When a drawer was open, Feynman found that the bolt twitched when he rested a finger on it as the combination dial passed the positions of the second and third numbers. That left him only twenty combinations to try. He got in the habit, whenever he was in an office with a drawer left open, of standing with his back to it. He found he could manipulate the dial, feeling for the twitch without looking, then glance back to note the combination position. The 'secure' units fell to his systematic determination.

As the bomb project neared its climax, Arline's health declined. Feynman drove over to Albuquerque pretty much every weekend and wrote a stream of letters but there was nothing to be done for her and she died in June 1945.

A few weeks later, Feynman was called back to Los Alamos with the message, 'The baby is expected.' Feynman returned in time to make the trip to Alamogordo, 320 kilometres (200 miles) to the south, to witness the first test. A 34-metre (110-foot) tower was constructed in the desert with a plutonium-based bomb at the top.

The test, codenamed Trinity, took place on 16 July 1945. The plutonium-based bomb sat out on the tower for thirteen hours while the team, located 9 kilometres (5½ miles) away in a concrete bunker, sat out a storm, terrified that a lightning strike would ruin their equipment. The concerns were unnecessary. The test went without a hitch. Another physicist present, Otto Frisch, said:

> That object on the horizon which looked like a small sun was still too bright to look at ... It was an awesome spectacle; anybody who has ever seen an atomic explosion will never forget it. And all in complete silence; the bang came minutes later, quite loud, though I had plugged my ears, and followed by a long rumble like heavy traffic very far away.

It was less than a month before bombs were dropped on Hiroshima and Nagasaki.

The end of the Manhattan Project must have filled Feynman with a mix of emotions. The project had been a success and a major factor in the ending of the war, but at the cost of many thousands of lives. And now Feynman could return to fundamental physics, but he had to cut his ties with the memory of Arline. In the end, he decided to accompany Hans Bethe, who was returning to Cornell University in New York State.

The first year was difficult, especially as Feynman's father died in 1946, but gradually the excitement of the chase returned. Building on his PhD thesis, Feynman helped develop the mind-bending theory that would gain him a share of the Nobel Prize – quantum electrodynamics, usually referred to for convenience as QED. The young British physicist Freeman Dyson said of it in its early form, 'It was a unifying principle that would either explain everything or explain nothing.'

As the Manhattan Project had demonstrated, modern physics was increasingly moving away from the solitary working practices once followed by the great names (although even Newton made significant use of the work of others, whether acknowledged or not). QED, which describes the interaction of light and matter at the quantum level, was not Feynman's work alone. The structure was built on the work of Paul Dirac, while two others effectively developed QED in different ways – Julian Schwinger and Sin-Itiro Tomonaga. These two shared the Nobel Prize along with Feynman.

The three differing methodologies were pulled together in a unified whole by another of Bethe's protégés, the same Freeman Dyson who had first seen QED balancing on an edge between everything and nothing. Each approach had its benefits, but arguably Feynman's was the easiest to grasp, using what would become universally known as Feynman diagrams.

The electromagnetic interactions QED describes account for the vast majority of the physics that we experience in everyday life. As Feynman himself once put it, 'QED is pretty much all of physics.' And in many ways QED was strikingly successful. Most physics theories use a model that produces something close to the observed result but QED came stunningly close to the measured values. It was as if, to paraphrase Feynman, you had a technique that predicted the distance from London to New York and the result was accurate to the width of a human hair.

QED's predictions were anything but intuitive. They described a quantum world where a particle takes every possible route (including taking a detour around the universe) to get from A to B. The way that Feynman made this mind-boggling concept accessible and usable was through his diagrams. Each particle on one of his diagrams could be accompanied by an arrow that rotated like a clock hand as time passed. The direction of the arrow showed a state of the particle called its phase, while its size showed the probability of the route being taken.

It sounds as if the approach was all pictures, but it still involved mathematics – Feynman used path integrals which took the sums over all possible paths. When the arrows of all particles on all the possible routes were combined, many cancelled out and the result was the expected one – that the particle travelled from A to B in a straight line.

If this were all that the method could do, it would be an extremely complicated way to produce a simple result. But it was not. It explained all the behaviour of light interacting with matter, such as reflection and refraction. It showed what happened when charged particles interacted. And, most impressively, it predicted that really strange things would happen in some circumstances – things that traditional physics would regard as impossible.

Take two quick examples. If you stand in a lit room and look at a glass window at night you see your own reflection. But someone standing outside can see you as a result of the light streaming through the glass. Most of the light passes through, but a small percentage reflects back. Not only does QED predict the chances of the reflection and of light passing through, it shows how the thickness of the glass can influence how many photons reflect from the glass and how many travel through.

Similarly, QED explains the way some surfaces (such as the playing surface of a CD) reflect strangely. If you imagine a photon of light coming in to a reflecting surface like a mirror and bouncing off, classical physics tells us it comes in on a straight line and reflects off at the same angle as it arrived (this is what is taught in basic high school physics). So you wouldn't expect to pick up the photons elsewhere. If you take the centre section of the mirror away, there should not be a reflection – and there isn't. But QED says that photons take every path, including travelling at a shallower angle, hitting a bit of mirror that's still there and reflecting back up at a steeper angle. You don't see these strangely reflecting photons because the different phases represented by the rotating arrows cancel out. But block off parts of the mirror, making it a refraction grating, and one set of phases can be eliminated. The result is that light reflects again, bouncing off at an unexpected angle. This happens in reality – as seen in the rainbow reflections from a CD that come from the little pits in its surface. These encode the sound and act as just such a refraction grating.

QED does far more than explain strange reflections. It gives a more realistic model for what happens when matter and light interact through an electron absorbing or emitting a photon – all represented by Feynman's elegant diagrams.

The approach was not without problems. It still had the habit of throwing up unwanted infinities, which after much soul-searching was dealt with by the process that Dirac hated, renormalisation (see page 220). The result was exquisitely accurate in its predictions, yet this renormalisation would never sit comfortably and remains an issue today.

With QED pretty much in the bag, Feynman was looking for a change of scene. Cornell seemed to have less regard for physics than for the arts and he found the weather depressingly cold. He toyed with South America but instead chose Caltech – the California Institute of Technology, based in Pasadena. At Cornell, Feynman had been something of a playboy but by the time he arrived in Pasadena he had a new wife, Mary Lou Bell, whom he met while she was a history major at Cornell.

Mary Lou seemed almost intentionally chosen as an opposite of Arline. Where Arline encouraged him to think his own way, to be his own man, Mary Lou wanted Feynman to conform to the rigid social structures of early 1950s academic life. She wanted a real professor – but ideally one who would not talk physics all the time, a subject she found less than interesting.

A clear contrast can be seen in their attitude to socialising. Feynman was a party animal who liked nothing better than an impromptu drumming session with a beer by his side and to be surrounded by bright young things with an interest in science. Mary Lou would rather he held his own at cocktail parties, making urbane remarks and preferably the only scientist in the room. An all-too-likely story told about Mary Lou relates how she was the only one home when Niels Bohr called in the hope of chatting with Feynman. Mary Lou sent Bohr away, saying Feynman was not available and later told her husband that she had saved him from an encounter with 'some old bore'.

It was around this time that Feynman switched his attention to the aspect of physics that brings quantum weirdness most directly into the macro world: low-temperature physics. At extreme low temperatures quantum effects are more directly observable and never more so than with superfluids – fluids that lose all viscosity and, once set in motion, are able to keep going for an indefinite time.

The first work on superfluids had been done back in the early years of the twentieth century, but it wasn't until the 1930s that mechanisms were developed to produce large quantities of liquid helium and the true nature of superfluids was established.

As we have seen in chapter nine, there are two broad types of quantum particle – bosons, like photons, and fermions, like electrons, differentiated by their spin values. Bosons are particles that can clump together in masses but there can be only one fermion in a particular state in any particular system. Generally speaking, individual 'matter particles' are fermions but linked together they can be a boson. So, for instance, the helium-4 isotope is a boson, because of the combination of spins in its constituent particles, while helium-3 is a fermion. However, at low temperatures, pairs of helium-3 atoms can link together to form a unified system and become bosons.

When such matter bosons are brought together at low temperatures they effectively act as a single linked entity, giving them strange properties in a substance known as a Bose-Einstein condensate. Feynman produced a stream of papers on the behaviour of supercooled liquid helium, using his diagrams to explain what was happening at the particle level.

By 1956, Feynman's marriage, which had been in difficulty for some time, had reached rock bottom and he and Mary Lou divorced. For a while he was wary of relationships. It was when he was visiting Switzerland that he met a young Englishwoman, Gweneth Howard, sixteen years his junior. He instantly got on with her but did not ask her out – instead he asked if she would come back to California to be his housekeeper. This was 1958, when such an offer would have been considered improper by many, but Gweneth was an independent and daring person who was technically on a world tour but had run out of money. Despite some concerns, she took the job, moving to Feynman's house in Altadena, California in June 1959.

By then, Feynman had returned to his habit of dating strings of different women. Gweneth largely lived her own life, though she did occasionally have an evening out with Feynman, which they both enjoyed. It came as quite a surprise, then, when Feynman proposed. They married the next year and stayed together until Feynman's death. Gweneth, it seems, was more a second Arline than a Mary Lou. She had the same strength of character which challenged Feynman to make the best of his abilities, whatever other people thought.

By this time, Feynman was deep into studying weak nuclear interaction, the mechanism by which a neutron in an atomic nucleus can split to produce a proton, giving off an electron and an anti-neutrino in a process known as beta decay (as a result transmuting the element as the number of protons increases). This is the process which Rutherford first isolated and named (see chapter six).

Feynman was by now an elder statesman among the new cohort of physicists, though he was still capable of questioning the accepted position to great effect. He pushed forward the understanding of the aspects of the nucleus controlled by the weak nuclear force and worked with Murray Gell-Mann to extend the principles of this beta decay to other particle processes.

What set Feynman apart, though, was his ability to think beyond his narrow field of expertise. Shortly before he married Gweneth, Feynman gave a lecture at the American Physical Society that is still regularly referred to by anyone writing about nanotechnology. In his talk, 'Plenty of room at the bottom', he argued that they would look back from the year 2000 and wonder why it took until 1960 to make progress in nanotechnology. In practice, we are still wondering, though some of the issues are understood better now.

Feynman proposed making small devices to build tiny machines that would then be used to make ultra-small technology, working downwards in scale while building more of these manipulators until it was possible to manipulate matter able to handle individual molecules on the scale of nanometres. We now know that the field is far more tricky than his suggestion made it sound because forces come into play on the very small scale that make traditional engineering designs unusable. We would need to use more 'wet engineering' – a field based on the same principles as biological mechanisms – but Feynman's vision still inspires, as did his offer of a thousand-dollar

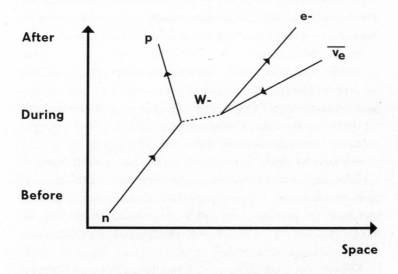

Figure 10: a Feynman diagram, devised by Richard Feynman to help in the calculations involved in quantum electrodynamics (QED). This diagram shows the decay of a neutron (n), via beta-decay, into a proton (p), an electron (e-) and an anti-neutrino. This particular radioactive decay is caused by the exchange of a W- boson, one of the bosons responsible for the weak nuclear force.

prize for the first person to produce a working electric motor less than 1/64th of an inch (0.4 millimetres) long – which was quickly won.

Similarly, after his work on the weak interaction, Feynman plunged for some time into molecular biology, a new area that recruited as many physicists as biologists. Here he worked in DNA, never quite making the right breakthroughs that soon after would lead to the cracking of the molecule's structure by Francis Crick and James Watson in Cambridge in the UK.

It was from the 1960s onwards that Feynman developed a reputation as a superb teacher, culminating in his lectures on physics. Many still regard these as the ultimate university-level introduction to the basics of physics. Yet, as always with Feynman, the reality was more complex than the myth. As American technology writer James Gleick notes, 'In fact, few physicists of even the middle ranks left behind such a small cadre of students or so assiduously shirked ordinary teaching duties.'

Feynman's reputation as a teacher was not in the conventional academic mode of nurturing graduate students and lecturing to undergraduates, but rather for his ability to spread the word to a far wider audience, whether through his popular books or the more technical Feynman lectures. (Even here the legend is strong – Feynman never strictly wrote a book. All the volumes with his name on the front are records of conversations or transcripts of lectures he gave.) Feynman was never interested in anything less than taking on the world. His popularisation paralleled the development of his own family life – with Gweneth he had the children he had always wanted.

By the mid-1980s, Feynman had contracted cancer which, combined with heart problems, left him physically weak. This was why it was Gweneth that persuaded him to take on the role that would expose him to the widest public attention he ever received. A role that brought him up against the might of NASA.

NASA's shuttle launches had become pretty much commonplace. Operational flights began in 1982 and the twenty-fifth mission in

January 1986 was really only of interest because among the crew of the *Challenger* was Christa McAuliffe, a teacher from Concord High School in New Hampshire. She was to be the first participant in NASA's Teacher in Space programme. Sadly, the mission took over the news headlines when, seventy-three seconds into the flight, *Challenger* broke apart, killing all seven crew members.

Feynman joined the commission set up to investigate the accident, invited by the acting head of NASA, William Graham, who had attended Feynman's lectures in the 1960s. Although Feynman had no experience with the space industry, Graham believed that his insight would be invaluable. This wasn't enough to get Feynman to take part, but Gweneth convinced him that his unconventional approach was exactly what the investigation needed.

Certainly Feynman's style contrasted markedly with that of the military personnel and civil servants who made up the bulk of the commission. He rapidly became frustrated by glacially slow progress and the carefully choreographed fact-finding visits. So Feynman made his own enquiries. Speaking directly with an engineer, he was given a lead to investigate the way that the large rubber O-rings used to seal joints in the rocket motor would respond to the freezing temperatures at the time of the shuttle launch.

Realising that vested interests were inclined to keep potential problems under wraps, Feynman took the opportunity to subvert a televised commission meeting that had been set up to regurgitate the blander findings made to date. With cameras rolling, Feynman took the floor, fixed a piece of O-ring rubber in a clamp and doused it in his glass of iced water. To a confused reaction from the commission, he withdrew the rubber under the watchful eye of the camera and removed the clamp. Rather than springing immediately back into shape, the rubber took seconds to recover, seconds that would have been deadly when under the pressures of active engines. Without its flexibility, the O-ring would not maintain a seal and the result was deadly failure.

Feynman spent the last ten years of life battling cancer and trying to get to the lost country of Tannu Tuva, a story dramatically portrayed in a BBC documentary made just a few weeks before his death on 15 February 1988. This documentary closes with Feynman and his friend Ralph Leyton playing bongos and Feynman making up a crazy song and laughing uncontrollably at the end. He was a showman and a joker to the last.

....

CHAPTER ELEVEN

Which list?

....

Now that we have met all the individuals involved, we can return to the list as presented by the *Observer* and rework it with members and rankings that more physicists would perhaps agree on. While it is difficult to see how any physicist could disagree with four of the names – Newton, Einstein, Galileo and Maxwell – the others are far more debatable.

Let's whittle down the possible additions. Many would want to include the quantum pioneers Werner Heisenberg and Erwin Schrödinger, whose matrix mechanics and wave mechanics (see chapter eight) tamed quantum physics and gave it a mechanism for making predictions, even if they brought in the probabilistic element that Einstein hated. Then there's the often-underrated Italian nuclear physicist Enrico Fermi. Not only did he construct the world's first nuclear reactor, Fermi predicted the existence of the neutrino and effectively discovered one of the fundamental forces of nature, the weak force.

Finally, we have Steven Weinberg's additions, Christiaan Huygens, a contemporary and near equal of Newton and Ludwig Boltzmann, whose application of statistics to physics, transforming thermodynamics and the way collections of atoms were analysed, is highly regarded (see chapter four). That leaves a list of too many, with a total of fifteen who are, in no particular order:

- Isaac Newton (1643–1727)
- Niels Bohr (1885–1962)

- Galileo Galilei (1564–1642)
- Albert Einstein (1879–1955)
- James Clerk Maxwell (1831–1879)
- Michael Faraday (1791–1867)
- Marie Curie (1867–1934)
- Richard Feynman (1918–1988)
- Ernest Rutherford (1871–1937)
- Paul Dirac (1902–1984)
- Werner Heisenberg (1901–1976)
- Erwin Schrödinger (1887–1961)
- Enrico Fermi (1901–1954)
- Christiaan Huygens (1629–1695)
- Ludwig Boltzmann (1844–1906)

Weinberg also commented that the list might be too Anglocentric, but we can't exclude people because of this – it has to be on their claim to a priority in pushing forward the boundaries of physics. For this reason, we have to take a step against diversity and remove Marie Curie. It would be wonderful to keep her, as female role models in physics are relatively few. And Curie deserved her Nobel Prizes. Yet her work arguably involved far more chemistry than physics and did not have the same kind of primary breakthrough in understanding as other top names.

We understand why Steven Weinberg wanted to include Huygens and Boltzmann but while there's no doubt they were impressive, their overall impact was quite narrow. That leaves us with a top twelve which we would rank as follows:

1   Isaac Newton (1643–1726)
2   Albert Einstein (1879–1955)
3   Galileo Galilei (1564–1642)
4   James Clerk Maxwell (1831–1879)
5   Niels Bohr (1885–1962)

6   Michael Faraday (1791–1867)

7   Ernest Rutherford (1871–1937)

8   Paul Dirac (1902–1984)

9   Erwin Schrödinger (1887–1961)

10 Enrico Fermi (1901–1954)

11 Werner Heisenberg (1901–1976)

12 Richard Feynman (1918–1988)

. . . and just drop the bottom two to get the top ten. Our suspicion is that the name most physicists would regret losing is Feynman but perhaps more from a kind of hero-worship than a real assessment of his contribution.

Considering how many individuals have won Nobel Prizes in physics since 1901, it's interesting that we have been able to ignore so many. In part this is because the pre-twentieth-century physicists effectively worked alone. They were essentially individuals, carving out their position for posterity. Since the 1920s the trend has been for work to be done in teams. At first, relatively small groupings came together but now, sometimes, vast scientific endeavours like the Large Hadron Collider at CERN have teams that rank in their hundreds. It becomes harder to pick out a key contributor to the development of our understanding of the universe. Even theoreticians are more likely now to collaborate than was once the case.

So what might this list look like in a hundred years? There is no doubt that we could have another Einstein – someone who looked at the problems we face in a totally different way. If an individual produced a whole new way of looking at the fundamentals of physics that, say, united gravity and the other forces – or replaced the big bang as the best accepted theory for the origins of the universe – then we would have a new contender. But, as yet, there is no one. An essential to be a Newton or Einstein or Galileo is that no one can see the need for your ideas beforehand. It's the change of direction in our thinking that makes these individuals so important.

Looking back at the list, there is another interesting development over time. All our early physicists had a strong religious belief. Of the more recent members of the list there are fewer for whom religion had any great significance. And this reflects a general trend across the scientific world, despite mathematics and physics counting more religious believers among them than other sciences.

In the original list, everyone up to and including Maxwell had a religious belief. The rest were mostly agnostic or atheist. Einstein had a tendency to refer to 'God' in his pithy remarks, but this was a code word that approximated to 'the organising tendency of the universe', rather than an independent deity. Bohr and Feynman were very open atheists. Dirac is an oddity in this respect, being vehemently anti-religious through to middle age, but embracing religion towards the end of his life.

Realistically, even our updated list will provoke debate and disagreement. You may well disagree with it. Feel free to share your thoughts on Twitter (@brianclegg and @RhEvans41) – we'd love to hear. The value of a list like this is considerably greater than the hundred best ever comedy programmes or ten best wines under ten pounds, because it provides a way to promote debate and to encourage us all to think about how one measures the worth of physicists. We believe that theirs is one of the most important jobs on the planet – and such a suggestion deserves careful consideration about what makes a physicist great.

We are, after all, talking about individuals whose expertise and knowledge, perhaps more than any others, gives them an understanding and mastery of the very fabric of the universe.

....

Index

·················